My secret EdTech diary

Looking at Educational Technology through a wider lens

Al Kingsley

JOHN CATT

First published 2021

by John Catt Educational Ltd,
15 Riduna Park, Station Road,
Melton, Woodbridge IP12 1QT

Tel: +44 (0) 1394 389850
Fax: +44 (0) 1394 386893
Email: enquiries@johncatt.com
Website: www.johncatt.com

ISBN: 978 1 913622 63 3

Set and designed by John Catt Educational Limited

For Niamh and Dan,
my proudest achievements.

Reviews

'Al is a pioneer in transforming school learning but has also been a consistent support for many of us working in the sector. He quietly champions the small fish whilst having a huge global impact. The nicest man in EdTech.'

Ben Whitaker, Director, EduFuturists

'Al is a great leader and thought provoker. He has incredible experience and insights and his heart is at the centre of education. He has supported and guided me on my journey throughout teaching and technology, and his knowledge and experience are invaluable to successful education with technology.'

Paul Tullock, Apple Professional Learning Specialist

'If you want advice about educational technology, from strategy support down to the implementation of one platform, Al is the man to go to. His friendly and honest advice has helped us help our teachers overcome many a challenge.'

Linda Parsons, Digital Lead

'Al is a shining light of the EdTech world. He is a fantastic advocate of EdTech collaboration between school leaders, education systems and the EdTech industry. Al champions and creates opportunities for teachers, school leaders and EdTech suppliers to work together constructively to share knowledge and best practice.'

Caroline Wright, Director General,
British Educational Suppliers Association

'Al is a breath of fresh air – constantly sharing learning, always offering a supportive hand and continuously pushing out his positivity!'

Drew Povey, Leadership Specialist

'Al is a brilliant educator and visionary in his views of EdTech. You can gain such strategic insight from him and his publications! It just improves your teaching all round.'

Caroline Keep, Maker Educator

'Al has been one of the most helpful, influential and supportive individuals in the global EdTech scene for several years now. He has not only championed those who have led the way on the embracing of EdTech in schools but has been a vocal champion for helping to find the digital superstars of the future!'

Abid Patel, Trust IT Director

'I love how Al encourages and inspires the next generation. My pupils have benefited so much over the years from his generosity, enthusiasm and opportunities offered.'

Martin Bailey, Digital Enrichment Leader

'Al is one of the warmest, most knowledgeable people in the EdTech world. He has a wealth of experience and graciously shares it with people from across the industry. His hard work and dedication benefit not only his own company but also young people across the country through his work as a chair of trusts and contributor to advisory bodies.'

Dave Leonard, Strategic IT Director

'Al is an inspirational man who is dedicated to supporting and challenging schools to reflect on how technology is used to motivate and enthuse children and staff. It is a privilege to work with him to continuously develop the IT provision at our school.'

Becky Waters, Headteacher

CONTENTS

Acknowledgements

I am enormously fortunate to work with an amazing group of people every day. They give me much of the motivation to push on, question and challenge and are always the source of plenty of smiles along the way. Being part of a team is so much more than just the tangible results – it shapes the journey and the memories you take away, and I have many. To everyone at NetSupport, Hampton Academies Trust, and all the other schools I work with – you rock!

I really enjoyed writing this book. It was quite cathartic to put so many ideas and experiences down on paper. In doing so, however, my wildly inconsistent use of punctuation, my obsession with capital letters, and my sometimes-challenging sentence composition would have sent many educators reaching for their red pens...or alcohol. So, a huge thank-you to Jackie Vernau for giving up her spare time to the thankless task of editing and proofreading all my work and applying more than a healthy sprinkle of grammatical sanity to my musings. Thank you so much, Jax (and Daisy, Buddy and Alfie)!

Alongside Jackie, a separate word of thanks to Eddie Elmore, who helped turn my charts from scribbles into a much more readable format – thank you, Ed!

Special thanks also needs to be given, of course, to my mum and dad, for giving me the education, skills and springboard to follow whatever path I wanted in life. It's taken a long time to fully appreciate the true value of that start. To Dave and Tricia, for trusting in me to steer our ship through both the calm and the stormy waters; and to Mrs Kingsley, for having to be both mum and dad far too often when my work-life balance was especially disproportionate. (In truth, it hasn't really changed much in 30 years and it definitely 'requires improvement'.)

Thank you to the team at John Catt for making this book happen. I'm very grateful to have had the opportunity to work with you all, and grateful in particular to Jonathan Woolgar for providing so much editorial support, enhancements and grammatical policing of the final edit.

Throughout the book and in my 'Voices Aligned' chapter, you'll see plenty of feedback, ideas and guidance from friends and peers across education. They, along with a brilliant online community, ensure that no matter what the question or challenge, there is always help and support at hand. As an online team, they are second to none, and I am very grateful and proud to be part of the community.

And as my final acknowledgement, I want to give an extra big thank-you and my utmost respect to the entire teaching profession for their adaptability, resilience and innovation in the face of the unprecedented situation and unbelievable expectations we faced during the COVID-19 pandemic; for leading the way, often without the support and appreciation of our national leaders (or, at times, our parents); for doing what needed to be done to keep our children, if not themselves, safe.

Alongside the teachers, whilst much of the content in this book is aligned towards the teaching profession, I am keen that our teaching assistants aren't excluded from the conversation or the recognition deserved for their efforts. And finally, I couldn't close the loop without also recognising that in the context of EdTech, it is the IT teams within schools throughout the country who suddenly had to become the technology experts, the new device prep and delivery facilitators and the infrastructure wizards to keep everyone connected. So, unsurprisingly, what we have is another perfect example of teamwork, with everyone having played their part.

Thank you, one and all.

Foreword

My eight grandchildren will leave school between 2024 and 2038. Let's take a moment to reflect on our changes in the use of technology, personally and professionally, over the last 12 months (who had shares in Zoom?), and now fast forward to 2038 and try to imagine what teaching, learning and assessment may look like. Now ask ourselves – what changes will schools and colleges have to make to their buildings and curriculum design, infrastructure, workforce development, leadership, governance, management, quality assurance and funding? And, last but not least, how will learning be assessed?

Trying to force new technology into old ways of working is not only futile but also counterproductive. It would fail to realise the true potential of technology to extend and enhance learning and engage and empower learners, and it would be a source of frustration for teachers and learners.

The design principles of our education system stem from a bygone era and were predicated on the dominant industries at the turn of the century – steel, coal, textiles, shipbuilding – and were strongly influenced by a form of Taylorism.

They are no longer fit for purpose in a digital age, and we need a generation of paradigm pioneers to rethink and redesign our education system. This means we must challenge the status quo, the outdated design principles of our system, something that I, and others, have been attempting for 25 years. It is an irony that it has taken a deadly pandemic less than 12 months to achieve what we have failed to do in all that time. But at such an enormous cost!

The COVID-19 pandemic has been a massive shock to the system – and we must learn from it.

The breathtaking irony was that after ten years of techno-scepticism at the heart of government education policy, the degradation of infrastructure and depletion of the education workforce capability and confidence in the use of technology, the government introduced a law to make online education a legal requirement.

The online genie is out of the bottle, and whilst there will be those who want to get it back in as soon as possible, teachers and learners have already decided that the future is bright, and the future of teaching and learning is blended.

Now we just need the archaic assessment system of high-stakes, end-test, written exams to recognise that the world has changed and that if they want to remain reliable, valid and, most importantly, relevant, they have an awful lot of catching up to do.

Professor Bob Harrison
University of Wolverhampton

An introduction to my EdTech 'diary'

Hi and welcome to *My Secret EdTech Diary*. I really hope you find this book a good catalyst to reflect on and plan how EdTech can play a role within your organisation. I should probably start with a confession – and in the context of this diary, it's a fairly significant one: it's not exactly a diary. It's a bit like a diary because it describes my 30-year EdTech journey, but it's not in chronological order and will contain relatively few salacious revelations. I've broken the book down into bite-size chapters covering different aspects of my EdTech journey, and each section ends with a selection of relevant articles I've written and shared in recent years.

If you are like me, I suspect opening a book on a topic like EdTech, with the anticipation of reading from front to back, is probably a big ask, and I'm not sure it's always the best way to consume the information. With that in mind, I've tried to write each chapter as a standalone entity, broken down into an introduction, the key strands of discussion, a conclusion and some supporting articles that summarise and highlight the takeaway points. Alongside that, throughout the diary (okay, this is the point where I should probably just call it a book) I have included, where appropriate, some easy checklists – in essence, questions to ask and thoughts to take away and reflect on that hopefully give you a good 'starter for ten'.

So, let's kick off with a fairly obvious first question: why a book on EdTech? There are plenty of resources out there covering all aspects of how education and technology have been interwoven for decades. I suppose it's because, for me, looking for the positives of 2020/21, EdTech has suddenly become significantly higher profile in the priority list of education discussions and, as a natural result, has also generated just as many questions as answers. What do we do? How do we do it? How do we pick the right solutions? How do we know they'll have an impact? How do we get started? The list of hows is extensive. Confession number two (which is slightly concerning given we're only on paragraph 3 of my book) is that I don't have all the answers – but I do have a few and I'm also in a position where I can share the successes and failures encountered by others during their EdTech journeys.

Another question I'm often asked concerns the different mindsets from a vendor perspective, as well as an educator one, when reviewing the role and value of educational technology. So, whilst we all probably subscribe to the general motto that 'the customer is always right', widening the lens through which we view EdTech requires us to also understand how vendors innovate new solutions, meet needs and help to shape the landscape for the future.

Please don't get me wrong, this doesn't mean that vendors know better or always get it right, because in fact a fair bit of what I'll be sharing reflects on the lessons that I believe EdTech vendors also need to learn. Nevertheless, if we can combine both to provide a broader view, then we will be better placed to make informed decisions – whether that be peer-, research- or experience-informed.

Those of you who have encountered me on your travels will also know that I have huge regard for my personal learning network (PLN). From the contacts and friendships I've developed through working in the space over the last 30 years, to the thriving online communities of Twitter et al., to the fond memories of meetings at educational trade shows around the globe, what I have learned and respect hugely is that unlike the commercial world where ideas are part of your intellectual property, in education the natural and overriding inclination is to share with your peers. As a result, much of what I've acquired in terms of experience has been built on that foundation. So I would be daft to attempt to write my first book – I mean 'diary' 🙂 – without also sharing advice, feedback and guidance from those trusted sources. You'll see quite a few mentions in the pages that follow.

Who is Al Kingsley?

So, I guess I should tell you a little about me? It seems only polite to at least give you a sense of my background and the experiences that have ultimately culminated in me putting pen to paper and creating this book. When describing my career, it's easier to split it into two neat and relatively equal halves, with both sides aligned rather nicely and which give me a fairly unique perspective on the successes, failures and potential of educational technology.

On the one side, I am group CEO of **NetSupport Ltd.** (www.NetSupportSoftware.com), an international EdTech software company which specialises in providing co-produced education solutions to help meet the needs of schools, trusts and districts around the world. Within this role spanning 30 years, we have developed a range of solutions which, like the layers of an onion (© Shrek 2001), support many aspects of school life.

The first layer starts with a solution that discovers, identifies and manages every device installed and used throughout a school, delivering a complete IT management platform which also includes eSafety and safeguarding features and empowers children to be able to report their concerns.

The next layer offers flexible and effective instructional technology that provides front-of-class teachers with control (and confidence) to both shepherd and engage digitally with their children, whilst simultaneously allowing for peer collaboration, quick assessments and a truly engaging experience.

Another layer within the technology portfolio offers a further range of solutions: creating observations, assessments and a 'learning journey' for our youngest learners; school alerting software to ensure messages are quickly received (particularly when considering school lockdowns); remote-control software to ensure the tech team can provide hands-on support to staff or childrens' devices no matter where they are; and a helpdesk to keep a log of all those requests for support.

All these solutions have evolved over the years because of co-creation with the education space, adapting to a changing landscape and, where appropriate, building a platform-agnostic approach to ensure they all work within a mixed IT environment.

So that's the business version of AI – 30 years of working with schools, ministries of education and school districts to develop solutions that are fit for purpose and deliver on their promises. It's been a real learning journey and one which has allowed me to delve deeper and understand the challenges (and opportunities) that technology can provide.

I did say there were two halves to me, and whilst education is an absolute passion of mine, and I grab every opportunity to contribute and add value to support the delivery of education in my community and the wider area, I also have almost 20 years' experience of school governance. At the time of writing this book, I am chair of two multi-academy trusts in the East of England, chair of an alternative provision academy and chair of the regional Special Educational Needs and Disabilities Board. As well as working with amazing and, frankly, inspiring people across those trusts, I am also very fortunate to be able to experience and work across the full spectrum of school settings. Our schools include infant, primary, secondary, and all-through provision, and also schools for children with complex needs and those who struggle with mainstream education and require a more tailored alternative provision.

With my experience supporting multi-academy trusts and my governance background, I also sit on the regional schools commissioner's headteachers board for the East of England and North-East London and in that capacity, I provide support for their work in developing and raising standards within academies across the region.

From my experience in school governance, I'm also chair of our regional governors' leadership group and for a number of years, I've hosted my own blog (www.schooltrustee.blog) where I try to share bite-size chunks of information, resources and best practice for the benefit of other volunteers working in school governance.

I am a firm believer that we are all lifelong learners, and I think it is important to support the broader pathways for young people as they leave mainstream education. So, alongside my trust roles, I'm also an apprenticeship ambassador for the East of England and chair of our regional employment and skills Board.

For the last year, I have been a council member for the Foundation for Educational Development (www.fed.education) and a strong supporter of their vision for a long-term education strategy for the UK, a fellow of the Royal Society for Arts and an active member and writer for the Forbes Technology Council (www.forbes.com).

As some of you will also know, I'm an avid article writer for a broad range of education titles around the world and an enthusiastic podcast contributor. Like many working within the education space, I'm only too happy to take part in discussions and debates concerning the effective use of tech, safeguarding and digital strategy (to name but a few of my favourite topics) and to learn from my peers and share ideas and advice on best practice. I can promise you, chatting on a podcast or panel is a whole lot easier than attempting to write a book – but if you're reading this, I must be on the right track.

In 2017, we set up **NetSupport Radio** with the help of the always engaging and enlightening Mr **Russell Prue** (www.AndertonTiger.com). This was really in response to the recognition that continuing professional development (CPD) and information sharing required people to have time – free time, that is – and for many, listening to interesting discussion and debate was often much more accessible while driving home from work, sitting on the train, or at their desk.

We took this a step further and brought our radio station over to the Bett show, broadcasting interviews live each day from the show floor. We had such a high level of engagement that we replicated this and took the radio station across to the UAE to broadcast live from the GESS show. Broadcasting from big annual education events has been a fantastic way of sharing some of the great conversations we have with attendees, and it has given us a platform to hear authentic student voices: plenty of our contributions come from children visiting the events. In 2020, just before the COVID-19 pandemic reached our shores, we aligned with the organisers and became the official broadcasters of the Bett show, providing a summary of daily highlights from the event and, hopefully, a useful conduit for those who were unable to attend. NetSupport Radio continues online and shares interviews and discussions with thought leaders from education, technology, and the broader community each and every month.

With the absence of those events (and, therefore, our ability to meet and share with our peers) during 2020 and 2021, we were also conscious of the fact that there were many new solutions being developed in response to the consequential challenges faced by educators, who themselves had ideas they wanted to share and take to market and who were struggling to find a voice and make their new solutions visible. With that in mind, my good friend **Mark Anderson** (@ICTEvangelist) and I decided to organise a one-off '**Check It Out**' show, where we could give those new start-ups some visibility and a chance to share what was special about their solutions or services. I recall that in the first show alone, we had over 15 different solutions, and we gave each contributor a five-minute slot to tell us about their solution, what made it different, and what they could offer to schools. Buoyed by the amazing feedback and interest in the format, we decided to do it again the following month. At the time of writing, it is perhaps no surprise to hear that these shows continue monthly, and the growing list of solutions and services suggests we might be doing this for quite some time. I really wanted to share this with you because it reflects the huge amount of innovation and creativity within the educational technology space and all of us – either as educators or vendors – have an obligation to nurture and foster innovative solutions and tools that continue to provide opportunities to raise the bar in our classrooms. You can watch or listen to previous episodes of the 'Check It Out' show here: www.bit.ly/316pOLO.

NetSupport

Check It Out Show

So that's me – a happy fusion of life in the commercial and education worlds. I'm also a firm believer in the big Cs, so expect a fair number of references to context, community, collaboration, communication, content, co-construction and creativity in this book, while I celebrate compassionate, capable and confident educators.

On a personal note, I'm also a dog lover – never underestimate the value of a furry friend for wellbeing (I have three lovely dogs that keep me grounded and motivated every morning) – and a classic car enthusiast. I combine my love of classic cars with the mindset that 'some things just don't need changing' and keep that in mind when talking about what we can do in schools because much of what we do and have been doing for years is already spot on – we should never embrace change simply for change's sake.

> Education is the most powerful weapon which you can use to change the world.
>
> NELSON MANDELA

Purpose

I think I've already confessed that I've never had any plans or aspirations to write a book, so it's not unreasonable to ask the question: why write one now?

Over the last 18 months, I've seen a huge number of educators sharing ideas and best practice and although the focus on topics is often very narrow, each has its own value and potential to support activities. I've learned a great deal from the abundance of information available, but I've also noticed a lack of information to help bring those silos of information together and provide more of a one-stop shop. To me, context is king, and under the mantra of 'sometimes you have to look backwards to move forwards', there is real value in having a much wider view of educational technology – past, present, and, of course, future.

So, what I have tried to do in this book is break the discussion down into some key areas. Section one is about unpicking the EdTech tapestry to provide some background, sharing a little history that potentially shapes our mindset and, on occasion, our preconceptions about the effective use of EdTech, and providing a useful glossary of keywords to help you navigate this often-confusing landscape.

Section two is all about lessons learned from the use of educational technology, looking at the various strands from the educator, leadership and governance perspectives and how it has shaped the areas of focus within a school where EdTech plays a role. So, think digital skills of staff and students, innovation, communication and, of course, wellbeing.

For section three, I figured it would be helpful to flip the perspective 180 and review the lessons learned, the considerations and driving thoughts that shape the solutions from a vendor's perspective. Although you may not be a vendor, understanding the challenges, expectations and thought processes a vendor has will hopefully provide the much-needed wider view for the end user.

In section four, I really try to focus on planning ahead – how the lessons learned and the process of reflection help shape the way we plan to best use and engage with technology moving forwards; and of course, within that (and not surprisingly for those who know me) I touch on the considerations for shaping a digital strategy.

I round off the book with reflections (or, as I call them, 'voices aligned') including feedback and great peer viewpoints on some of the key questions raised in the book, before wrapping things up with a few of my favourite top 10s – from blogs, podcasts and events to, of course, books.

With every step we take, where relevant, I've also included copies of dedicated articles I've recently written which focus on that particular strand, just to add some additional context.

Whilst I don't presume this book will provide you with all the answers, what I do hope is that it will save you precious time by consolidating key information and resources into a common place and, most importantly, get you thinking, get you questioning and, I dearly hope, get you engaged with the potential of educational technology. Enjoy!

As I like to say: 'When appropriate, and used effectively, EdTech can truly be the jam in the learning sandwich.' It doesn't have quite the gravitas of a quote by Nelson Mandela, but I like it!

Unpicking EdTech

So, what is EdTech?

Given that the word is on the front cover of this book, it's incumbent on me to define what I mean by EdTech or, to use the extended phrase, 'educational technology'. In the world of Twitter, you will generally find it linked to the hashtags #EdTech and #EduTech. I'm no shortcut researcher so of course I referred immediately to the source of all knowledge – Wikipedia – to seek out a suitably concise and eloquent definition, but in truth I ended up immediately wanting to change the scope of their definition. Wikipedia neatly defines educational technology as 'the combined use of computer hardware, software and educational theory and practice to facilitate learning'. At first, it seems a fair encapsulation of the term, but the author was, perhaps, viewing it through too narrow a lens. Let me expand...

When we talk about educational technology, our natural inclination is to consider it solely within the context of the learning environment – i.e. the classroom (or more recently, the online classroom) – but actually, I believe educational technology should be considered in a much wider context as it represents the effective use of technology in *any* part of the educational setting – whether it's to support your school, your trust, or your district. It's fair to acknowledge that 90% of the narrative around educational technology relates to supporting effective teaching and learning; however, we also know that the right technology, when deployed more widely, can improve operational efficiency and communication which, in turn, can free up valuable time and (with the right tools) precious money to go back into our classrooms. So, when we think about the wider deployment of digital technology within our schools (see my later chapter for lots more thoughts and ideas), the more we can save and the more efficient we can become in the broader operation of our schools, the more flexibility and funds we have available to focus on our core priority – teaching and learning.

So, whilst the original definition isn't wrong, for the purposes of this book, let's try to keep a much wider view. Perhaps we can redefine educational technology, especially when planning for the future, as 'any computer hardware, software, device or service intended for deployment within an educational setting'. After all, it would be easy to exclude a WiFi access point from the original definition, yet it's pretty key to the use of tablets in the library, for example. You may agree with me that mine is a rather broad definition (and it is) but in reality, the role technology plays within the operation of any school is huge and covers so many strands, it's counterintuitive to try to define it as one single thing. Nor, for that matter, is it likely to be manageable by the skillset of one individual. This is actually a key point that needs to be made – it's easy to compartmentalise the use of technology into a very narrow subset and assume a mindset of 'geeks can't teach, and teachers can't geek'. In fact, what we quickly realise is that there's a huge role to play for both teachers and IT technicians – and we need both around the table to make the right decisions.

Does having a suitable definition help us much? Well, in one sense, no – because it simply highlights the fact that we are talking about a very broad and diverse subject.

On the other hand, yes – if it opens our eyes and ensures we don't fall into the trap of adopting an overly narrow viewpoint while we consider the broader topics covered in this book and the resulting questions we absolutely should be asking.

One thing you quickly learn when working within education is that there is a term for everything – and most of those terms are abbreviated (and abbreviated again), just to make it difficult to remember what everything means 😉

To help, below are a few of the terms often referenced, which sit one layer below our EdTech definition and encompass different aspects of school life where EdTech plays a pivotal role.

Some are common, some less so, but if it saves you having to put your hand up in a meeting for clarification, I'll take that as a win.

CAD	computer-aided design
CAI	computer-aided instruction
CAL	computer-assisted learning
CBI	computer-based instruction
CBT	computer-based training
CLAIT	computer literacy and information technology
CSCL	computer-supported collaborative learning
DTP	desktop publishing
IBT	internet-based training
ICT	information and communications technology
ILS	integrated learning system
IT	information technology
MALL	mobile assisted language learning
MIS	management information system
MUVE	multi-user virtual environments
SIMS	school information management system
SIS	student information system
TEL	technology-enhanced learning

VLE	virtual learning environment
WBT	web-based training

I was browsing through my Twitter feed one morning and I saw a really good question posed by **Shane Guildford** (@shanelegend23). He asked peers for their best 'You should use EdTech because...' analogies, and his own suggestion really encapsulates, for me, the mindset I am trying to foster and what I hope we can share together throughout this book:

> Do you run or exercise? Yes. Do you run or exercise better listening to music? Yes. Well, think of teaching as running and you'll do it much better thinking of technology as the music.

I really like this analogy (I guess if I didn't, it wouldn't be in the book 🙂) but one of the key messages I try to share is that EdTech plays the role of the facilitator – something that underpins all the amazing activities that occur within a school – and so I think the analogy of music helping people get the most out of their own natural talents is a really good way of explaining the role of technology. To round off this point, and to address the small minority who might still have the wrong end of the stick, technology is not there to replace teachers; used appropriately, it's there to help teachers and students and to support staff.

Perhaps I should wrap up this topic with a quote from **George Couros** (@gcouros), author of The Innovator's Mindset (Couros, 2015) who simply states:

> Technology will never replace great teachers, but technology in the hands of a great teacher can be transformational.

And another quote I really like comes from **Youki Terada** (@YoukiTerada – Terada, 2020), who states:

> Good technology integration isn't about using the fanciest tool; it's about being aware of the range of options and picking the right strategy – or strategies – for the lesson at hand.

A wider view

Okay, so hopefully at this point all the cogs are starting to whirr and we're thinking about educational technology in the broadest sense or, as I like to say, through a wider lens. Hopefully too, we've got a pretty solid understanding of what falls under the umbrella of EdTech and that the starting point is a mindset that recognises that, when used correctly and appropriately, EdTech can have a positive impact on life within a school.

Without wishing to labour the point, many of my reflections (successes or failures) about our response during the COVID-19 pandemic, and all the strands we need to consider when planning a digital strategy, require a broader view than merely considering the classroom. I think this is a good time to mention a pretty well-known and -evidenced model, **SAMR** (Puentedura, 2010), and stretch its boundaries a bit. I'll cover the model in more detail later in the book when we pick up on the teaching and learning aspects, but for now, and in a very small nutshell, SAMR stands for **substitution, augmentation, modification and redefinition**. It categorises four different degrees of classroom technology integration. I mention it here because I think the same rules and considerations apply on a broader scale.

As an example, let's take school parents' evenings. It has been shown, rather successfully, that these can be substituted (and augmented) by an online parents' evening system which, for many, is easier to attend, more private and, for better or worse (depending on which side of the table you are sitting), results in better time management. The definition of 'the classroom' just got a bit blurrier. In addition, our go-to meeting platforms of Zoom, Microsoft Teams and Google Meet augmented the approach to school governing body meetings – we swapped the classroom for an online discussion and added the value of a recorded session for those who couldn't attend to supplement the usual meeting minutes. We are shifting further away from the classroom and the original intention of Puentedura with the SAMR model, but to me (and hopefully you too) it makes just as much sense applying it to a wider view.

History

In my cogitations whilst planning this book, I was mindful that we needed to look back before we can move forward. It's actually a cornerstone practice of a good digital strategy (see my 'Planning Ahead' chapter) and builds on the reflective nature of educators. Where I struggle though is how, in the broader annals of an education timeline, looking backwards serves any significant purpose other than, in many cases, reminding us of the pitfalls of the past...and missed opportunities.

When I started researching the history section of my book (and I'll be honest, 'historical research' is probably an overstatement), the first real tech I encountered was back in the 1920s with Sydney Pressey's teaching machines. Just the name 'teaching machines' immediately indicates there was perhaps a subtle exaggeration of the capabilities of the solution. Mr Pressey was an educational psychology professor at Ohio State University who developed a machine to help provide 'drill and practice' items to students in their courses. (Pressey, 1926, p.374). The summary quote Pressey used for the mechanical device was 'Lift from her [the teacher's] shoulders as much as possible of this burden and make her free for those inspirational and thought-stimulating activities which are, presumably, the real function of the teacher.' It sounds not a million miles away from the narrative we hear now with how AI (artificial intelligence) tools can help support personalised learning and retrieval practice. So, 100 years on and in some respects, we are still seeking the same outcomes. Having dug a bit deeper, the solution was akin to a typewriter with four keys so, in my simple head, it was the first iteration of the all-conquering multiple-choice test!

I don't want to understate the achievements of the aforementioned, but wishing to keep this book light-hearted and accessible, it did make me smile to read about a device with just four buttons for interaction which was developed by a man called Pressey. No? Okay, perhaps just me then.

The next stop on my brief educational research journey led me to the 1950s and the work of a B.F. Skinner who, after developing the experimental analysis of behaviour, wrote *The Technology of Teaching* in 1968, which became one of the first narratives around technology and its use in teaching.

B.F. Skinner created the 'programmed instruction' educational model (PI). He first introduced this with J.G. Holland at Harvard University by self-administered instruction which was self-paced and presented in a logical sequence with multiple content repetitions. Skinner argued that learning could be accomplished more easily if the content was divided into smaller incremental steps and, just as importantly, learners got immediate feedback, reinforcement and reward. There were two models of programmed instruction. The first was linear where, as indicated, content was divided into small steps (always the same steps) and where learners could go at their own pace and immediately see the results. The second model, referred to as the branching (or intrinsic) model, was introduced by Norman Crowder. In this model, students had to address the situation or problem through a set of alternative answers. If they answered correctly, they moved on to the next step; if they got it wrong, they were detoured to remedial study depending on the nature of their mistake. I may be guilty of confirmation bias at this point, but it seems we've all been thinking along the same lines and trying to achieve the same things for quite some time.

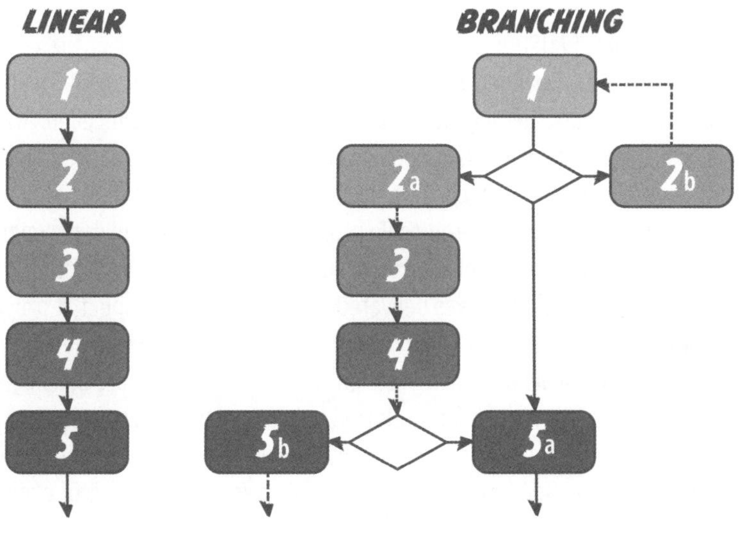

To conclude this micro-thread, in the 1960s, an Indiana University psychology professor, Douglas Ellson, took a look at PI, identified its weaknesses and developed a new process referred to as programmed tutoring (PT) which, quite simply, added the human factor (the tutor) (Ellson, Barber, Engle and Kampwerth, 1965). Just like PI, students worked at their own pace reading and solving problems, but with PT, the tutor watched and listened and when the student hit a roadblock, the tutor provided hints, took them back to something they already knew and helped them to move forward again. During the 1980s and onwards, this approach gained credibility due to its track record in comparative studies (Cohen et al., 1982) and was recognised by the US Department of Education as one of the topmost successful innovations.

This was a bit of a segue but certainly served a purpose for me – it's a reminder that few of the approaches we reflect on are entirely new; we are almost always building on the shoulders of others and evolving the way we undertake tasks.

Of course, not all projects have quite the same longevity or positive impact as others. How about an easy one – roll the clock back just a little way to 2013 when Los Angeles Unified School District spent $30 million on an iPad contract with Apple.

It was a big deal – LAUSD was the second-largest school district in the country and the project was designed to revolutionise many aspects of digital learning. However, fast forward to August 2014 and the project was cancelled, with a headline message that it was a failure of vision, not the technology. I'm not going to unpick this example in detail because it's clear there were challenges at every step of the way for such a new and large-scale project – from successful deployment of the devices to the integration of the curriculum software that they wanted to use, and the planning and CPD that was delivered alongside. In one respect, it provides a useful lesson that technology alone cannot deliver a solution – you need to plan strategically for this scale of implementation. In truth, it's much more of a negative because at least once a year I hear it referred to in a presentation by someone, and sadly it only serves to raise extra caution and cynicism about the potential impact technology can have. Don't get me wrong, I'm all up for a healthy bit of cynicism;

but the devil is always in the detail and as in this case, often all that people hear or remember is 'the big iPad project that failed' when in truth, many of the issues were much more on the human level.

I'm taking a meandering journey in this section, because whilst it's easy to get dragged into the detail of specific solutions or inventions over the years, most of the narrative around the use of EdTech is linked to projects that provided the magic key – funding, which has often been the catalyst for a sudden surge in uptake and new EdTech acquisition and, of course (just like LAUSD), presents the opportunity for a 'spend quick and plan later' approach.

One such example that springs to mind, and for quite a few positive reasons, was in 2006 in the UK when the government announced a new 'Building Schools for the Future' (BSF) programme for secondary schools across England. The programme was delivered by partnership schools and was a joint venture between the Department for Education (DfE) and the private sector. The first wave included engagement from 14 local education authorities, and by 2009, there were a total of 96 in the programme. Alongside the BSF programme for secondary schools, in 2007 they also added the 'Primary Capital Programme' (PCP) with just under an additional £2 billion available to spend on 650+ primary school projects. I'll explain why this is in my history section shortly, but as with all government initiatives, certainty rarely remains for long and by 2010, the Secretary of State for Education, Michael Gove, announced that, following a review, the programme was to be scrapped. By June 2010, 178 schools were rebuilt or refurbished with a further 231 under construction.

The reason why I've included the BSF and PCP programmes in my reflections is because yet again it highlights the need for long-term planning and stability to make a significant and sustained impact in our schools; and when decisions about investment and potential outcome are made within a single parliamentary term, they need to have long-term benefit. That said, at the time, BSF was what we needed to free up available capital and deliver a new wave of modern school buildings to support the education system. In parallel (and key to this book), as well as providing the capital for the fabric of each school, it also funded the delivery of

ICT and technology (and CPD) within those schools. In fact, there were a number of schools who achieved funding exclusively under the 'ICT-only schools' category, although they only totalled about 20. At the time, the overriding need was for investment and funding to expand our school estate.

There were lots of positives wrapped around the BSF programme, and in addition to the obvious one of getting the new schools built, it also shone a light on the ICT opportunities available and, for many schools, provided a welcome boost and fit-out. I'm keen to avoid too much negativity, but as with many front-loaded ICT projects which form part of a new build, the real challenges come further down the line – realising that the ongoing school capital funding doesn't continue to support an adequate refresh cycle. So, we start off with well-resourced and -equipped ICT labs and classrooms but sadly, over time, devices fail, we don't have a full complement of devices for a whole class, and we absolutely don't have the budget to do a full refresh and renew on a typical four- to five-year life cycle. Translate that into plain English – if you want to sustain ICT in the classroom but don't have reliable, up-to-date equipment, nor the confidence in its likelihood to perform as you expect, then within a relatively short period of time, you will find its role, adoption and use gradually diminish.

Ten years on, and I can see a similar pattern emerging once again with the current 'free schools' programme which was introduced by the coalition government in 2010. At the time of writing, 'wave 14' projects have been agreed in principle; and on a personal level, I've been involved in the delivery of both a secondary and primary school into a trust under the free school programme. In the early days, the free school programme had a bumpy ride, not least because any group had the opportunity to apply based on the need to build a school within an area where there was demand. Unfortunately, post-build, in some cases the demand just wasn't substantiated, the schools weren't full or effectively managed and, as a result, a small number regrettably proved to be a poor use of public money. That changed some years ago and the recent wave of free schools has, for the most part, delivered some really good new schools into areas where there was a clear need for extra capacity and choice.

I fully support the free school programme in the broadest sense (given current options), but I'm also keen to dispel the misconceptions that they somehow operate in a different way, outside of the normal academy structure, as this is simply not the case. For all those on a learning journey as I was, a free school is simply a different route to securing funding to build your school. Once built, they are simply an extension of the academies programme (Academies Act 2010).

'Interesting stuff, Al, but why have you included free schools in this section?' Well, hopefully you are beginning to see that these things tend to follow the path of evolution rather than revolution; and as I've mentioned with the upfront funding allocation for IT within a new free school, although it's healthy to cover the phased opening of your new school, the exact same problems arise further down the line when it comes to funding the renewal and refresh of the technology – it really is a *Groundhog Day* experience from BSF and previous initiatives.

For anything to become evidenced and embedded in a balanced and reflective way (especially within education) so that it delivers tangible impact, it has to be consistently sustained over time. So, if we build a new school and we're able to install appropriate ICT resources in all the dedicated areas, use them to deliver effective teaching and learning whilst at the same time imperilled by the sword of Damocles – 'We'll never be able to afford to do this again' – hanging over our heads, the likelihood is that over time, the impact will inevitably diminish.

Take a typical secondary school in England, with circa 1500 children. The upfront ICT funding for a new free school ranges from £300k–£400k, which of course includes a lot of switches and servers and WiFi infrastructure alongside the obvious computer hardware and software. It sounds like a reasonable pot of money but, as always, it doesn't go very far and only provides a basic level of functionality. Roll the clock forward a few years and consider that the capital funding allocation for that typical secondary school is likely to be £17k–£20k per annum (www.bit.ly/33iJNO3) and you can start to see that the odds of a school being able to maintain that initial level of technology is virtually impossible. From that £17k–20k, they also have to maintain the building and everything within it – including the shiny, new IT – so, to have a decent refresh cycle, it has to be funded by taking money from the school's revenue budget and that's just not practical for many schools. So, once again, I come back to that 'long-term' vision – it's tough to plan if you don't know what your funding is going to be from one year to the next.

Please understand, this is not me having a direct pop at the education system in England (although I'm sure we all agree more funding would be helpful right now), because if we head back over the Atlantic, in the same way as we can discuss the pros and cons of free schools, there is a parallel narrative taking place about the Charter Schools Program (www.bit.ly/2RsHd4Q) and how effective it has been with public funding of those entities. I won't broaden the conversation to go into more details about charter schools but, regrettably, there are examples within the programme that highlight where a lack of confidence in the effective use of technology continues to arise. One good example of this is from Ohio – the Electronic Classroom of Tomorrow (ECOT) claimed to serve 15,000 students as part of its virtual delivery, but state officials found those numbers were significantly overstated and ECOT was forced to close in 2018, owing $80 million in funding. I haven't seen or heard anything to suggest that the technology involved or delivery of their curriculum was at fault, but it doesn't really matter in terms of the narrative – what the history books reflect is that the big projects are often linked to technology that failed. The true operational, strategic or resource reasons for failure are never included in the postmortem.

If I tackle the same narrative from a different angle, I could reflect on the story of Amplify, a new company launched by Rupert Murdoch's media company, News Corp, in 2012 which promised to deliver a completely digital classroom. It was built with technology procured through the acquisition of Wireless Generation, an organisation which already provided technology to schools, including the engine for NYC's department of education's School of One maths programme. Anyway, in 2013 Amplify launched its new tablet with lots of fanfare, yet by the end of that year, one of its largest customers, Guilford County Schools in North Carolina, announced it was recalling all the devices in use due to technical issues, including 1500 broken screens and melting chargers. By 2015, Amplify and their completely digital classroom were put up for sale and subsequently sold, having failed to deliver on their promise.

At about the same time that Amplify was hitting the news waves as our future digital classroom, a brand even more well-known than Rupert Murdoch came up with their own technology announcement – again in 2012, Google launched Project Glass, their 'very polished and well-designed wraparound glasses with a clear information display above the eye' (Bilton, 2012). Within weeks, education technology blogs were aligning themselves with the education sector, identifying all sorts of ways the glasses could be used – Adam Bellow even delivered the closing keynote speech at ISTE 2013 wearing a pair (Bellow, 2013). I read some of the education media at the time, including a list of 30 ways Google Glass 'might' work in education, composed by online education provider Open Colleges, and to describe a number of the suggestions as tenuous would be (generously) giving the benefit of the doubt (Pajaron, 2013). Well, you decide – wearing Google Glass, you could 'make teacher evaluations, removing the observer from the physical classroom'. Yup, that's the future: have a teacher wearing glasses so they can be observed and evaluated in the background through what they were seeing. It's a bit too Big Brother for me, and I'm not expecting many hands up in support of that suggestion. In practice, what this showed is when the tech is 'exciting', there is a real risk that we try to create and align solutions to problems that don't even exist, just to indirectly validate the need for the technology. That's just never going to fly.

Aside from all the privacy and security concerns around Project Glass, in 2015 Google announced they would stop selling the product. But at the time of writing, this area of tech is back on the radar as the next generation of interactive data consumption. It's déjà vu all over again! (I'm not sure how many times I'll have to use that phrase in this section – sorry!)

I might be at risk of opening a whole new can of worms with this one, but what about interactive whiteboards (IWB)? In 1990, Xerox Parc introduced the first interactive whiteboards to be used in meetings and soon after, in 1991, Smart Technologies introduced their first interactive whiteboard. In 2007, the British Educational Communications and Technology Agency (BECTA), as part of their 'technology schools survey', indicated that 98% of secondary and 100% of primary schools had interactive whiteboards and a year later, in 2008, the interactive whiteboard industry was valued at over $1 billion a year.

Even now, with most new-build schools, one of the first items on the shopping list is an interactive whiteboard at the front of the classroom – but you surely have to question their real use and impact, don't you? I'm asking because clearly some still don't question them and purchase IWBs just because that's what they already have in other classrooms. In my humble experience working in and visiting many schools, an interactive whiteboard is often just used as the surface on which to write and annotate – much of the supporting functionality within the whiteboard rarely sees the light of day. Perhaps this is a good example of how the money was spent on the shiny and new, without setting aside a suitable pot to fund the CPD that goes alongside the technology. I look at the modern classroom now and think that a large-screen TV would have greater impact alongside appropriate screen-sharing technology. For the last 15 years, it's been the ability to 'project' that has been the key part of the equation, not the surface that's being projected onto. In a positive way, the debate continues and is increasing, but who knows how much has been spent on resources like this without significant challenge or consideration of alternatives.

Wearing my vendor hat, I could add to this section the joy of Windows RT, introduced by Microsoft in – you guessed it – 2012. That certainly was a year for innovation 😉. In a nutshell, Windows RT was a cutdown or lightweight version of Windows 8, built for the 32-bit ARM processor. Microsoft hoped this would provide a platform that was simpler, easier to manage and delivered longer battery life. In practice, everyone was familiar with Windows 8 and immediately applied comparisons with RT and the incredibly limited software ecosystem available given that applications could only be accessed via the relatively new Windows store at the time. As a vendor, my personal experience of the engagement with Microsoft and the time spent, under their encouragement, migrating applications to work on Windows RT was not a good one. Let's just say we spent an inordinate amount of time and money supporting their requests, but RT was gone without any warning and before there was any opportunity of a return. In this case, it was absolutely the limited ecosystem and functionality that was the nail in the coffin of Windows RT and by late 2014, it was all but dead (Ranger, 2015).

I'll expand on this further in my vendors' perspective section because the actual concept of lightweight, easy-to-use devices, aligned with a centralised ecosystem for apps and low management overhead to compete with the Apple and Android tablets, of course makes perfect sense – and Google proved that to be the case with the Chromebook.

There is a common theme here: projects or products which have failed, particularly the high-profile ones, are memorable and sow the seeds of doubt; those that are successful tend to continue under the radar with little or no media interest. So, having looked back and now planning to look forward, much of this section is about highlighting the fact that, alongside investigating and evidencing how technology can be used within your organisation, you also must be mindful of the fact (partly justified) that there will always be scepticism and caution about the true value of any proposition.

We also need to look outside our immediate sphere of influence and learn lessons from others. In 2014, the Further Education Learning Technology Action Group (FELTAG) published a paper entitled 'Paths forward to a digital future for further education and skills' (FELTAG, 2014), which was the result of research by the Institute of Prospective Technological studies. The paper highlighted that the FE sector must keep abreast of change, as digital technologies will have a 'profound effect on the economic and social well-being of the country' and 'learners must be empowered to fully exploit their own understanding of, and familiarity with, digital technology for their own learning'. It even made a recommendation of delivering 10% of teaching online by 2015. Change the narrative from further education to secondary and there are clearly lessons we could have learned if we had connected the dots.

What have we (hopefully) learned?

- Don't always believe the hype about a new solution – look for the evidence.
- Lack of consistent funding has been at the heart of holding back sustained use of tech in schools.
- Don't plan without considering whether you can sustain what you introduce.
- Big brands don't always guarantee the safest investment.

- Much of what we consider 'new' isn't actually new – it's an evolution of something else, so be informed.
- Speak to your peers for evidence of success – the media will likely only showcase the failures.

Let's end on a positive from my brief (and intentionally quite selective) history. In recent years we have seen tablets (Android and iOS) arrive as affordable additions to the school ecosystem; Chromebooks too have had a very positive impact; lots of new software solutions have evolved that have made a difference; augmented reality (AR) and virtual reality (VR) solutions are becoming more mature; the cloud has unlocked the opportunity for accessing software as a service (SaaS); and almost every child has a phone in their hand nowadays and, therefore, the potential for it to be used as a learning platform.

Data has become much more accessible and, when correctly deployed, more centralised. In recent years, the rise of data aggregators and accessible reporting platforms and dashboards have made decision making and evidencing impact easier – whether from centralised progress and attainment monitoring or whole-school virtual SOAP ('school on a page') reports – using tools like Microsoft's Power BI (http://powerBI.Microsoft.com). As we will cover later in the book, communication and parental engagement are key considerations, and there has been a significant rise in technologies to share childrens' work, behaviour and associated updates with parents, direct to their smartphones – tools like ClassDojo, Edmodo, ReallySchool and SchoolStatus as some examples.

There are lots of positives right now. We've just got to make sure we make the most of the opportunities presented to us and think about shaping our own digital strategy...which is convenient, as I cover that later in the book.

Glossary

This section explains some frequently used terms that may assist you on your learning journey. Apologies that it's a bit of a mixed bag – but isn't that always the case in education?

1:1	When a school issues a laptop or tablet to every enrolled student
AAC	Augmentative and alternative communication – any communication method that helps individuals with speech and language impairments to communicate
Active Directory (AD)	A database and set of services developed by Microsoft that connect users with the network resources they need to get their work done. The database (or directory) contains critical information about your environment, including what users and computers there are and who's allowed to do what.
adaptive learning	Software that adapts its content and pacing to the current knowledge and skill levels of the user
asynchronous learning	Teaching and learning that leverages online resources in such a way that students and teachers do not need to be in the same place at the same time
AUP	Acceptable use policy – policy providing guidance on the appropriate use of devices within the school
AWS	Amazon Web Services (their cloud offering)
Azure	Microsoft's web services (their cloud offering)
blended learning	A method of teaching that combines online and face-to-face activities. Students usually have some control over time, place, and pace. (Post-COVID-19, it can also refer to a mixture of asynchronous and synchronous learning.)
bring your own device (#BYOD)	A model that allows, and even encourages, students to use their personal phones, laptops, or tablets to assist with classroom instruction (Can be a real pain for the school IT team to manage)
Clicker	A device or mobile app that allows students to answer a multiple-choice question
Common Core Standards (CCS)	The set of US learning standards
course management system (CMS)	A broad term for systems that keep both students and staff organised with all their digital resources, discussions, assignments and schedule management
device agnostic	In simple terms, solutions that work across the full range of devices in a school (Windows, Mac, ChromeOS, iOS and Android)

differentiated instruction	Adjusting and adapting instructional methodology, programmes and resources to address the learning styles and ability levels of individual students
digital citizenship #DigCit	Teaching children to navigate the digital world safely, responsibly and ethically
digital literacy	The skills for using technology competently, including interpreting and understanding digital content and creating, researching, and communicating with appropriate tools
flipped classroom	'Flipping the classroom' requires students to view pre-recorded content (i.e. the material that would traditionally go into a lesson) for homework, and then to devote class time to discussion or group work activities.
gamification	A strategy for engaging students more deeply in learning through game design principles and mechanics
hybrid classroom	A combination of physical face-to-face and online teaching and learning
learning management system (LMS)	A platform for the administration, documentation, delivery, tracking, and reporting of online learning activities
learning platform	A learning platform is an integrated set of interactive online services that provide teachers, learners, parents and others involved in education with information, tools and resources to support and enhance educational delivery and management.
MOOC	A massive online open course
PAAS	Platform as a service – an environment for developers and companies to create, host and deploy applications (e.g. Azure, AWS, etc.)
perpetual licence	Pay for it once and you are licensed to use it forever – but typically licenses will also have an annual cost to keep the software up to date and supported.
personal learning network (PLN)	An informal network of professional people meant to aid in furthering an individual's skills and experience
personalised learning	Tailoring lessons, assignments and assessments to individual students
personally identifiable information (PII)	Any data or information collected by an academic institution or partner that can be traced back to a specific student

project-based learning (PBL)	A teaching method based on 'learning by doing' – students work hands-on on a real-world activity that demonstrates the concepts they are learning
SaaS	Being a bit lively and full of spirit...okay, it also stands for software as a service – online-hosted applications and services you typically pay for on an ongoing subscription basis.
screen time	Any time children spend in front of televisions, computers, phones, video games, or other electronic devices
STEAM	STEM plus the arts
STEM	An acronym that stands for the fields of science, technology, engineering and mathematics
synchronous learning	Real-time teaching and learning where instructors and learners participate in instant two-way communication
the cloud	Nothing mysterious here: it's just a term to describe a group of servers and applications hosted remotely on someone else's infrastructure, which is accessible from anywhere.
virtual learning environment (VLE)	An online learning system that supports full access to, and management of, classes, assignments, resources, tests and more

Connecting the dots

We often reflect on the broader topic of EdTech and how it influences and impacts school life for all our learners from age 4 through to 18, and that's a great starting point. However, there is a need to reflect on how that then flows into the skills our children have that can be transferred into the working environment. It's not an unreasonable point to consider whether the tools we select and use to educate our children will still be of value as they transition into working adults. Given the pace of technological change, it's something I regularly try to encourage school leaders to factor into the mix.

Finally, to wrap up this discussion, in April 2019, I wrote an article for Innovate My School (www.innovatemyschool.com) where I posed the question: 'The future tech-driven workplace: is it all down to EdTech?' I've shared a copy of the article below.

INNOVATE MY SCHOOL – APRIL 2019

The future tech-driven workplace: is it all down to EdTech?

Much has been written about the future workplace. What will it look like? What sort of jobs will there be? Will there still be a 9–5 office culture? Will automation mean fewer jobs?

We can speculate about artificial intelligence, green power, wearables, data-driven, and virtual environments and so on, but the truth is, nobody knows exactly what the workplace will be like by the time today's five-year-olds get there. That leaves education sector professionals in the position of having to ensure the young people in their classes are equipped with a comprehensive range of digital skills as they progress through school – not to mention the creative abilities they'll need to innovate and problem solve.

Students have been using technology in school for some years now, amassing a wide range of digital know-how, and this is supported by their increasing use of personal devices at home. So, it's not really surprising that there are already whispers that the young people of Generation Z, who are now entering employment, are shocked to find that workplace technology isn't more advanced than it actually is.

Tech awareness starts early

It's been just over a decade since the iPhone was released in 2007 and the first Android smartphone appeared in 2008. Since then, tech has become completely woven into the fabric of everyone's lives. And it's been a rapid change. Think about it: we now do our banking, bill paying, appointment making, socialising, travel ticket buying and more online – and we control our homes and entertainment choices with smart gadgets and digital assistants. As it's now almost the norm for young people to have their own smartphones, little wonder then that toddlers are pressing and swiping on devices before they can even speak.

Schools have had to act quickly to support this. Not only do they have to ensure that enough time and money is invested in training in order to maximise the use of the solutions they buy, but they also need to meet the requirement to teach in line with the DfE's Digital Skills Framework's five categories of core competencies that will prepare students for every area of life and work: communication, information handling, transacting, problem solving and being safe and legal online.

Many are doing this in brilliantly creative ways, incorporating the use of digital tools into every subject rather than just teaching computing in isolation – and the students are totally absorbed, with many achieving things with technology that leave their parents' generation not even knowing where to start.

Every bit of exposure to EdTech prepares students for the future, even though it's a given that technologies will change significantly before they reach the workplace.

What effect will missing workplace tech have?

Our changing society will largely inform what the future workplace will be like. Technology has given us the means to do many things as soon as we think of them – and that's something we've fast become used to. We can look things up on Google and get answers straight away and we can talk to our friends en masse instantly via social media. In a nutshell, we've become used to doing what we want, right now.

That's what today's children will expect when they get to the workplace. Being asked to use outdated systems or technology that hinders their progress just isn't going to wash. Wider tech developments have been enormously freeing, giving people the flexibility to choose when and where they work and the ability to access data from any device, anywhere, at any time. So, if there is a restriction in the path of something they know can be done faster in another way, at its simplest, it will mean they'll either install the tool themselves or do it at home where it's available to them. After all, they've been taught to be creative problem solvers at school...

Put yourself in their shoes and it's easy to see that the frustration of constant tech obstacles such as these could well be a trigger for young people to change jobs – and it's something that's already being talked about.

Lessons for the future

Could it be that increased use of EdTech is driving changes in employment technology? I think we're going to hear more and more about this over the next few years. EdTech is certainly helping to fuel the pace of technological change that we've seen to date across all sectors – not just in terms of the application of students' digital skills, but also by prompting increased innovation by school IT solution developers as we respond to meet the demands of the EdTech sector.

The rapid growth of and access to personal technology is also playing its part, with this resulting amalgamation of personal and school digital knowledge contributing to our young people knowing instinctively what is – or should be – possible with the ideal tools to hand.

Swift technological change has placed new demands on everyone, especially employers. In general, workplace technology tends to lag slightly behind developments in these other areas due to cost and efficiencies (the thinking being, 'Why rush to change something that already works?'), so it could be that, increasingly, frustrations from the younger end of the workforce will force businesses to consider technology not only as a process improvement mechanism, but as a retention tool; after all, there's no point in succeeding in attracting top talent if they're going to leave in three months for an alternative post where there's the tech in place to help them be at their most productive.

It's important to remember that technology won't be the answer to everything in the future workplace, just as it won't be in education. Tech solutions used in either sector should be carefully considered for their proven impact and sustainability. We'll still need human skills and attributes such as creative problem solving, face-to-face communication, empathy and the adaptability to relate to each other. However, the combination of this with thoroughly embedded tech knowledge and digital skills will make for an increasingly productive workforce. There are exciting times ahead.

LESSONS LEARNED

When I first started writing this section, my immediate thought (or, more appropriately, dread) was that this had the potential to develop from a chapter in my book into a collection of views and resources sufficient to fill something akin to the full *Encyclopaedia Britannica*, and that most definitely would not be in line with my desire for focused, accessible information.

What I've tried to do is condense my ideas, what I've seen, what I've read, and what I've learned from my peers across the educational landscape into concise topics that I believe reflect our primary area of focus.

I should start with a preamble shaped very much around the changes in students' needs and in teaching and learning approaches – after all, there's little point if they aren't at the heart of the discussion – but, as with the introduction, before we delve into this further, I also wanted to flip the viewpoint around, so we maintain the broadest possible perspective throughout the book. Let me explain.

If I look outside of the education space to the financial markets and the investors' litmus test of where they see the 'educational technology' sector evolving...well, let's just say the narrative is overwhelmingly optimistic, with a boom in EdTech investment.

Although we know that spending money on technology has no immediate correlation with good decisions, effective impact or outcomes, the investment market is usually a pretty accurate reflection of the level of appetite and focus a particular area has – in other words, the view is that if the vultures are circling, you know there's an appetite for action. (The comparison to vultures is perhaps rather unfair in the sense that for many of the solutions we use in our schools every day, external investors have been the positive catalyst for development and delivery.)

Let's just briefly put that market context into numbers. In March 2021, Researchandmarkets.com released the 'Europe EdTech and Smart Classroom Market Forecast to 2027', under which headline they forecast that the European EdTech and smart classroom market is expected to reach a value of $6.1 billion by 2027 (from a value of $2.5 billion in 2019). They are forecasting a compound annual growth rate of almost 15% per annum. Their overall summary was as follows:

> Despite the negative impact of the COVID-19 outbreak on economies of EU countries, the demand for EdTech and smart classroom solutions has seen an upward trend owing to increasing demand for e-learning, virtual classrooms, and other digital technology solutions for uninterrupted delivery of education to students across Europe. The sudden adjustments in education systems and processes in the wake of the pandemic is pushing stakeholders to invest in modern technologies and adapt to the evolving technology landscape in the education sector.

I've included the markets and numbers here as part of my introduction to this chapter simply to validate a sudden and, frankly, seismic shift from what many of us have seen in the education space for years – namely, that there is finally an appetite and focus on how educational technology can support our schools. Now of course an appetite has to be backed up with funding, and funding requires a change at the very top in terms of priorities and politics, but the markets are normally 'on point' with their analytics. So, before we delve into this chapter, we can at least hopefully be on the same (positive) page – recognising there is now a clear and undeniable spotlight on EdTech.

Of course, we should all still be cautious. Many of us have been through the *Groundhog Day* process of initiatives that spark the potential for investment and push forward, only to be short-lived with another change of direction. Somehow, post-COVID-19, it does feel like now, more than ever, there is an opportunity to genuinely make a difference. In order to make that difference, we need to learn our lessons: we need to know what's worked well and, more importantly, what hasn't, so we don't just keep doing more of the same. Equally, as we return to a sense of normality within the education landscape, there is a real danger that the positive lessons we've learned and the new skills and tools that have been adopted fail to be embedded and, as a result, fail to be part of our onward journey.

What a terrible waste and missed opportunity that would be.

Many of the themes and processes described in that brief history highlight a very design-driven system. In a 2020 presentation entitled 'Post-Covid-19 – A blended future for Teaching, Learning and Assessment', **Professor Bob Harrison, University of Wolverhampton** (@bobharrisonedu) shared (and referenced in this book's foreword) how most of our systems and the design principles are predicated on a process born of the industrial revolution. The thought processes were very mechanistic, and the education system was built around the same approaches and design principles used in the industries of that time. These processes are no longer fit for purpose and the innovations we have seen with teaching, learning and technology in recent years are *despite* that legacy system, not because of it. We need a paradigm shift in the way we think about teaching and learning rather than trying to force new technology into older working methods.

> The drivers of the education system – assessment, curriculum, inspection/quality requirements, funding flows, promotion criteria – have not changed in recognition of what technology offers, so nothing within it can change.
>
> DIANA LAURILLARD

Lessons learned: teaching and learning

In many ways, this is the most important section of my EdTech diary – and when I say 'in many ways', I mean in almost *all* ways; teaching and learning is the fundamental driver for what we do. I appreciate that in the previous chapter I advocate that the definition of 'educational technology' should be broadened to ensure we deliberate over *all* aspects of technology across the school estate, but pedagogy is absolutely at the heart of everything.

If I roll the clock back to 2015, I must confess that my optimism in the sector had waned; it really felt like we had atrophied our focus down to a few core technologies that weren't, in most cases, particularly well research informed. With so many other priorities, there really wasn't an appetite at senior leadership level to look for innovation outside, nor was there any analogous desire to fundamentally review the way we did things. We have a brilliant leadership team within my trust – some of the best people you could ask for – but I will confess to having personally endured the frustration of highlighting the need to reflect on our digital strategy and getting nods of agreement, along with an ever-growing list of convincing reasons why there really wasn't time right now. The reasons why not were genuine, but sometimes the priority just has to shift and, hopefully, in an upward direction. It took a good 18 months to finally get some focus on a topic that I appreciate, for many on our senior leadership team (SLT), was neither in their comfort zone nor something they could really pull positive experiences from to generate a catalyst to prioritise.

I guess that perception also manifested outside, albeit with the caveat of possible confirmation bias, with us not really seeing any significant changes within the broader landscape of EdTech events during that period. On the whole, I'm a big supporter of the key EdTech events around the world – notably **Bett** (www.bettshow.com), **ISTE** (www.ISTE.org) and **GESS** (www.gessdubai.com) – but it was hard not to notice that when we looked at innovation and future technologies – particularly the start-up sector grouped at Bett, for example, under the umbrella of 'Bett futures' – they were typically tucked away in a secluded corner and not the real driver at those events. A little bit too much of the same old, same old was emerging.

Moving through to 2019, I'd argue there wasn't any compelling change, aside from some emerging technologies particularly around topics like augmented reality (#AR) and virtual reality (#VR) that were becoming a discussion point. The one significant change was in the hardware landscape where Chromebooks continued to become more prevalent, largely based on their price point and ease of central management. Perhaps I could throw into the mix the discussions around visualisers becoming more popular front of class instead of interactive whiteboards, but I must be brutal and say there were no other real significant shifts in terms of the way we did things.

Of course, during that period there were lots of new individual solutions that arrived and added value, and we did plenty at NetSupport (www.NetSupportSoftware.com) that I'm very proud of, but for the purposes of this chapter, I'm focusing on the big seismic changes in the narrative.

So, welcome to 2020 and, courtesy of a pesky little virus, suddenly the narrative and focus changed – literally overnight.

Necessity is the mother of invention.

Teachers had to suddenly and magically produce remote teaching (and learning) solutions – pulling the proverbial rabbit out of the hat – in a matter of days. For the next 12 months, instructions (okay, orders) would arrive from the government either at close of play on a Friday afternoon or, helpfully, on a Sunday...just to help support wellbeing I suppose 😉. The teaching profession became, in military terms, a bit of a rapid reaction force. It's no surprise that if you try to do something on a massive scale with little or no notice, it's going to take time to get it right. I don't do politics, but over the coming months it certainly felt a bit hypocritical that schools were expected to deliver a perfectly blended model online to an expected standard, yet it was equally fine for a track-and-trace system to stumble along for months on end with massive investment and little to show for it.

I won't digress any further or I'm liable to get a little grumpy, but I suspect those reading this book may have a viewpoint in common with my own on this topic.

So, there we were, throughout the world, around March 2020 (some a bit sooner, some a bit later), having to figure out what would work for our schools. For most schools, I think the first challenge (which became a realisation very quickly) was the scale and scope of what we now refer to as 'the digital divide' – those children who had access to appropriate technology at home during the school day and those who didn't.

Most schools quickly undertook surveys with their parents and children to try to get a clearer picture of the landscape and then had to add into the mix of considerations those households with more than one child and the associated challenges of concurrent use during the day. Hindsight is a wonderful thing, but I hold my hands up to the fact that I never thought to ask, pre-COVID-19, whether we had checked if all our children had access to appropriate technology to do their homework on a regular basis. Perhaps it was just felt that we could simply measure based on output and those without would have been visible to their form tutors, but even limited access can have a significant impact on the time and quality of learning and, for older learners, a significant impact on their ability to research.

After rapidly figuring out who had access to suitable technology, the next challenge was the inequality in terms of connectivity and associated performance. Not all families have access to reliable broadband, and as the redundancies and furlough schemes kicked in, we saw families' budgets contract and, in some cases, mobile data plans disappear. As we learned over the coming months, some solutions have a much heavier data footprint than others, and that had a significant impact for some households. The solution for many schools was to make 4G dongles available for targeted households, but even that plan had limitations – with the explosion of use online by all households (for work, education, and recreation), localised data speeds and services became stretched.

In parallel, schools also had to undertake the same process to identify equipment and connectivity available to staff, many of whom were delivering their teaching remotely. For some located in a rural area, it was clear that connectivity was key, and confidence in an online lesson was certainly underpinned by a reliable connection. For many, technology and the physical hardware in their hands was the most pressing issue. It was one of those times when plenty of schools, including some of mine, wished they had had the opportunity to wind the clock back two years and know what they knew now when making decisions about teaching equipment. Hindsight again, eh?

As is often the case (and, I would advocate, almost always wrongly), decisions about the best technology to deliver teaching were made under the considerations of affordability, manageability, and budget restrictions. That distilled down to a quick comparison which identified that laptops were more expensive than desktops and needed more repairs, which made desktops more popular and presented the added value that there was always a suitable piece of equipment front of class, even if the lesson was being covered by someone else. It kind of made sense until we started talking about the delivery of remote teaching, and then suddenly, the onus was thrust onto some staff to have their own personal equipment at home that would allow them to continue working.

Of course, that inevitably resulted in a whole raft of questions, including device suitability, availability of a webcam and microphone, appropriate apps and curriculum resources being installed alongside security, privacy and antivirus considerations. Many learned a very quick lesson here – the schools that were able to adapt the most quickly and successfully were the ones where staff had their own dedicated device provided by the school and, as a result, had a higher level of confidence using it during lessons.

Let's put the tech challenges to one side for a moment – I don't expect what I've said above will be a revelation to anyone – and agree that the bigger question everyone had to get to grips with came from a pedagogical perspective. Schools weren't solely trying to create a 'handout and collect' homework style – they were also factoring in how to instruct, question, challenge, support and give feedback, all the way through to nuances like pacing the flow of information as they would with their observations in the classroom.

Due to the regional timings of the pandemic, I saw many of the international schools in the Middle East move to online delivery first. This was also facilitated by the fact that their cohorts all had access to suitable technology from the outset. This was the catalyst for the now familiar discussions about Zoom, Microsoft Teams and Google Meet. These platforms came to the fore as the most accessible ways of delivering an online experience. Ironically, many schools were already Microsoft- or Google-enabled and had access to these tools, but they were largely dormant in their use prior to the pandemic, and as a result there had been negligible CPD on them.

You are Muted

Along with the adoption of those platforms, a whole raft of new familiar phrases became synonymous with our collective learning journey: the frequent 'You are muted' reminders during an online session; the quick check 'Is that a legacy hand?' when we forgot to put our virtual hand back down after raising a question; the regular discussions about the appropriateness of the virtual background while online. All these elements formed part of the impetus to create suitable policies and guidance for the delivery of online lessons and, alongside that, best practice and safeguarding considerations. As with most things along the journey, they were all part of a domino effect triggered by the original national announcements.

With Zoom and Teams and our new 'virtual classrooms', procedural, policy-driven guidance was key in striving for the best chance of success. I've seen lots of great examples of suitable guidance, but for the purpose of this chapter, it can be distilled down into two summaries.

For students (and their parents/carers)

Sharing key tips, advice and expectations with students and parents is key to ensuring they know how to use the most suitable platform appropriately and have an opportunity to become familiar with and adept at using it. Alongside that, a best practice guide would include the following:

- Parents should review how to use the platform with their children, so they all know what is possible and permissible (e.g. muting sound, using chat, screen share etc.).
- Parents should attend for the first few lessons with younger children.
- Use calendaring tools to map out session schedules.
- Check cameras and microphones are working and set to the right levels.

- Make sure necessary equipment is nearby, e.g. books, pens and other stationery.
- Wear appropriate dress when on camera.
- Choose appropriate backgrounds.
- Engage in video sessions within appropriate spaces.
- Close other browser tabs to reduce the load on computers.
- Keep devices plugged in to mains power.
- Keep pets and other distractions away.
- Emphasise appropriate behaviour.
- Remember lessons may be recorded for safeguarding reasons.

Like all CPD, I'd argue that it shouldn't be a one-off – even if it feels well embedded, it's good to be mindful of in-year admissions, as well as new intake, each academic year and ensuring there is an ongoing programme of advice and guidance.

For staff

In parallel, for many teachers, this has been an equally challenging digital voyage of discovery, so sharing similar guides an best practice became just as important, with CPD on each solution being top of the list.

A best practice guide for teachers delivering online lessons looks something like this:

- Have consistent processes for all teaching staff, share those processes with teachers and, as with an AUP, have a similar policy in place for remote learning.
- Don't use personal devices or accounts.
- Wear appropriate dress.
- Be mindful of the background, even if you can use green screen type effects.
- Have a few practice runs with colleagues to ensure they are comfortable.
- Use available tools (e.g. those within Google and Microsoft) to set up lessons/schedules.

- Engage in video sessions in an appropriate location.
- Close other tabs, programs or apps to reduce device load.
- Keep pets and other distractions away.
- Share a calendar of planned video sessions with students.
- Think about body language.
- If possible, record sessions and save to the cloud.
- Check desktops and other open programs – when screen sharing, ensure personal items aren't visible and close other apps or tools that may contain personal information.

With all the above in mind, the next discussion became much more subjective – or needed to be anyway – and surrounded schools' interpretation of 'online learning' and how best that should be delivered. Some of the very earliest adopters felt that the gold standard was to mirror a physical classroom experience with an online schedule – or, as we like to describe it, become 100% synchronous. Some of this pressure was driven and amplified by parents and their expectations, rather than being based on empirical evidence.

Pedagogy trumps the medium.

On the subject of evidence, in April 2020 the Education Endowment Foundation (EEF) (www.educationendowmentfoundation.org.uk), published a report entitled 'Remote Schooling: new EEF evidence review highlights core features that can unlock its potential'. Perhaps unsurprisingly, they found that the *quality* of remote teaching was more important than *how* lessons were delivered. They gave examples of teachers explaining new ideas in a live session versus a pre-recorded video and highlighted that what was key was whether the explanations built on a child's prior learning, not how the session was delivered.

There were other useful highlights within the report which identified not only the challenges surrounding access to technology but also the importance of finding ways for pupils to interact with each other within the online learning environment – e.g. through peer assessment – and how that could boost the impact of remote learning. The report summarised the five key findings as follows:

- Teaching quality was more important than how lessons were delivered.
- Ensuring access to technology was key, especially for disadvantaged pupils.
- Peer interactions provided motivation and improved learning outcomes.
- Supporting pupils to work independently improved learning outcomes.
- Different approaches to remote learning suited different types of content as well as pupils.

Within the focus on teaching quality, the suggestions (which in practice are no different to those within the physical classroom setting) identified the need for clear explanations, scaffolding and feedback. At this point, in my efforts to reflect the narrative and best practice around this topic, I make no apologies for deviating from technology to pedagogy for a couple of pages – it's very relevant.

In his article 'Reflecting on remote teaching: Exemplifying several strategies' (published in the EEF Rapid Evidence Assessment Report in March 2021 by HISP), Steve Smith, director of HISP Research School (www.researchschool.org.uk), shared the key strategies they adopted to support 'high quality' live teaching, which were as follows:

- **Planning the lesson** using the home learning planning framework adapted from the EEF's 'Metacognition and self-regulated learning guidance' report.
- **Starting the lesson** with a slide (five minutes before lessons start) setting out information to allow students to organise their workspace and remove distractions such as phones, and to reinforce expectations for the use of microphones and the chat function.
- **Activating prior learning** by starting each lesson with a form of retrieval practice such as a short five- to ten-minute quiz through Microsoft Forms. Other strategies included building a slide deck of key images from previous lessons and asking students to elaborate on them by typing into the chat

function, reading aloud the responses, making corrections, and addressing any misconceptions.

- **Linking the lesson objective** to previous lessons using a slide to exemplify the 'learning journey so far'.
- **Teaching new concepts** by utilising an electronic graphic board (with a laptop and electronic pen) to give 'live worked' examples, with a focus on delivering clear, high-quality explanations. Explanations were followed by a learning check (quiz or discussion) and answers discussed with a set of relevant images.
- **Using assessment** to check deeper understanding through one or two 'key application' questions per topic, answered 'open book'. Students could send a photo or electronic document of their work via email. This offered an opportunity to discuss responses during the next lesson. Pupils were asked ahead of time if they were happy to share their work and were assured that this would either help to address a misconception or highlight a good explanation.

Following that initial list of key drivers from the Education Endowment Foundation and having considered the access to technology and challenges surrounding delivery of the lessons, the next focus was on fostering peer interactions.

It's easy for some of this narrative to get dragged upwards into the domain of the secondary landscape (or years 7–12 in the US), but there were good strategies being adopted in primary schools, from EYFS to year 6, that were, for some, just as effective. Again, in many cases the choice of tool – such as SeeSaw (www.seesaw.me), ReallySchool (www.reallyschool.com), Tapestry (www.tapestryjournal.com) etc. – was secondary to the strategies applied. A good example of creative approaches to promote greater peer interaction was shared by Kate Atkins, head at Rosendale Primary School (www.rosendale.cc). She separated strategies adopted for EYFS and KS1 and KS2 as follows:

For early years:

- Small group, live lessons to mirror the Kagan cooperative learning approach at Rosendale. These small group sessions used a round robin structure to offer opportunities for oral rehearsal with particular language structures ('If I were an animal, I would be a ________ because...') or counting skills.

- When practising number skills in live lessons, teachers calling out a number and asking the children to demonstrate that number by balancing on different body parts on the floor – for example, 1 elbow and 2 knees if the teacher called out the number 3.
- Setting up a live Lego play session so that children could play together side by side. Some pupils told each other about their construction, while others chose to play quietly.
- Teachers sharing photos of children's home learning activities with the aim of replicating the interactive, idea-exchanging environment of free play.

For key stages 1 and 2:

- Teachers posing questions in their whole-class morning message, asking pupils to find the answer in time to explain during live lessons.
- Teachers using examples of pupils' work to support others in verbalising their metacognitive thinking – for example, sharing how a pupil had used language successfully in a piece of writing.
- During live teaching, teachers using Kagan structures 'Take Off Touch Down' and 'Find the Fiction' – for example, a teacher would say 'Take off if you have a dog' and pupils would see who else is a dog owner.
- Teachers taking their year group on a virtual school trip to Antarctica, which included a memory game, a pre-flight quiz to gain a visa and a scavenger hunt. This was used to support pupils' writing.

The penultimate strand of the EEF's findings centred on how outcomes could be improved by supporting children to work more independently. They cited multiple reviews that identified the value of strategies which help pupils work independently with greater success. Wider evidence related to metacognition and self-regulation suggested that disadvantaged pupils were likely to particularly benefit from explicit support to help them work independently – for example, by providing checklists or daily plans.

I've always advocated the 'one size doesn't fit all' approach when considering children, or even schools, and the idea of following a very prescriptive approach for all seems flawed given that every teacher will know their own cohort best and, therefore, how they can maximise engagement. It was good to see an acknowledgment within the EEF's fifth finding that different approaches to remote learning suit different types of content and students.

It shouldn't be a surprise to anyone that when it came to the challenges of delivering online learning, technology and the right application platform provided the infrastructure needed, but the human-to-human element is what made the lessons and determined the outcomes. We all hopefully appreciate that the 'right' tech can support and encourage engagement, interactivity and save time in stewarding a session, but as I always say, #EdTech is just the facilitator of great teaching and learning.

When I talk about the balance between technology and teaching during presentations, one slide I often include showcases the technological pedagogical content knowledge (TPACK) framework (Mishra and Koehler, 2006).

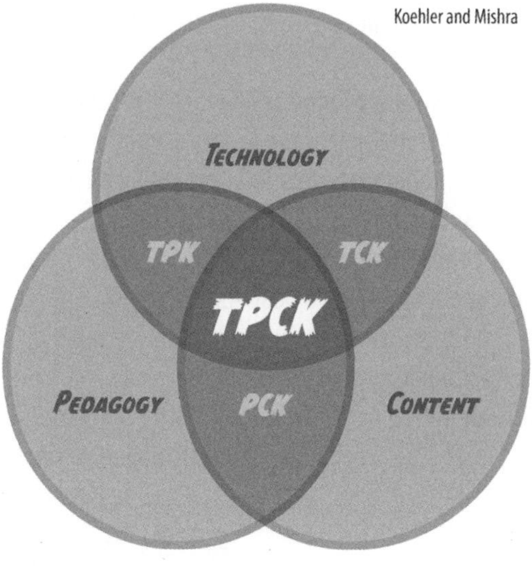

For many I am sure this isn't new, but for those who haven't encountered it before, at the heart of the TPACK framework is the interplay of three primary forms of knowledge:

- **Content knowledge (CK)** – 'Knowledge about the subject matter to be learned or taught.'
- **Pedagogical knowledge (PK)** – 'Understanding how students learn, general classroom management skills, lesson planning, and student assessment.'
- **Technology knowledge (TK)** – 'Knowledge about ways of thinking about, and working with, technology, tools and resources.'

On reflection, I should probably have turned the diagram upside down – it works better (for me) with the technology underpinning or sitting below the main order of the day. TPACK wasn't always on my radar; I'd heard it mentioned a few times but didn't really give it much thought. Then I watched a Mark Anderson (@ICTEvangelist) presentation where he referenced it and that was the catalyst to review and reflect on how it fits within the EdTech landscape. What I like most about displaying the concept as a Venn diagram is that it visually highlights the fact that technology has a role to play, *when appropriate* – it doesn't always need to be part of the equation. As you can see, there are multiple interactions between the three areas in the diagram. The smaller sections are described as follows:

- **Pedagogical content knowledge (PCK)** – 'The teacher interprets the subject matter, finds multiple ways to represent it, and adapts and tailors the instructional materials to alternative conceptions and students' prior knowledge.'
- **Technological content knowledge (TCK)** – 'Teachers need to understand which specific technologies are best suited for addressing subject-matter learning in their domains and how the content dictates or perhaps even changes the technology – or vice versa.'
- **Technological pedagogical knowledge (TPK)** – 'An understanding of how teaching and learning can change when particular technologies are used in particular ways.'

And then finally, when all skills align together, in the centre you have:

- **Technological pedagogical content knowledge (TPACK)** – 'The basis of effective teaching with technology, requiring an understanding of the representation of concepts using technologies; pedagogical techniques that use technologies in constructive ways to teach content.'

In 2019, I was fortunate enough to be invited to present at the JESS (Jumeirah English Speaking School) Digital Innovation Summit (www.jess.sch.ae) in Dubai. While I was at the event, I watched a presentation by **Kate Jones** (@KateJones_teach) who referenced the same Venn diagram but added an extra ring around the outside – to reflect context. I'm not sure who originally added it to the concept, but it made sense to me and hopefully we all agree, context is key in so many ways.

Connecting the dots, it was that combination of technology underpinning the pedagogy and content knowledge that had been fast-tracked, using the SAMR principles mentioned in chapter 1 (Puentedura, 2010), to substitute some of our physical resources for digital equivalents and, in the case of using Teams and Zoom et al., to further augment and modify the way teaching was delivered and content was accessed.

The 2019 report 'Mediating factors that influence the technology integration practices of teacher educators' (Nelson et al.), which was based on 800+ US teachers, highlighted the positive link between TPACK and alignment with ISTE standards, identifying it as a strong predictor of alignment. Its recommendation was that schools should provide targeted support to educators across disciplines and should adopt coherent technology frameworks.

Calm before the 2nd Storm

That brief detour brings us to an important juncture in our EdTech journey – a place I think most schools arrived at after navigating the first lockdown; a reflection and consolidation period prior to the second wave.

I'm going to park the challenges of managing bubbles, concerns about distancing within classrooms and all the other variables schools had to deal with because I really want to focus on the period where we reflected on how the first lockdown had gone, the lessons we had learned and the contingencies that needed to be considered in advance of any future lockdowns. The reality check was the lack of time which had been allocated to prepare and deliver sufficient CPD so that staff were able to confidently use the technology and, of course, the cornucopia of supporting apps we had hastily adopted for delivering online learning.

Of course, hindsight is a wonderful thing, but at the time, many of us were hoping a return to predominantly on-premises teaching and learning was likely; the second wave (and threat of more) soon threw that hope out of the window.

The UAE was one of the first countries in the eye of the storm and early in 2020, as a collaborative initiative between the UAE Ministry of Education, the Abu Dhabi Department of Education and Knowledge, the Sharjah Private Education Authority and KHDA, they provided a framework for evaluating and reviewing the status of online delivery within their independent schools. The evaluation was carried out through discussion with school leaders, review of students' work, observation of online lessons, and feedback from parents, students and teachers. It was a relatively simple model that focused on three key 'zones' – students' distance learning and wellbeing, teaching and monitoring students' learning, and leading and managing students' learning – and was presented in a grid for review, as shown below. Each section was assessed and identified as developed, partially developed, or not developed.

Zone A	Zone B	Zone C
Students' distance learning and wellbeing	**Teaching and monitoring students' learning**	**Leading and managing students' learning**
Attendance and participation	Planning and delivery	Agility
Safeguarding	Sharing intended learning outcomes	Contingency
Learning opportunities	Distance learning programme	Communication and engagement

Equity of access	Monitoring and assessing learning	Resources management
Wellbeing		

Although relatively simplistic, I still think it usefully highlights some of key areas of consideration for delivery (and in a later chapter where we discuss shaping a plan for a future, the above strands are worth bearing in mind). There wasn't a comparable exemplar to work with in the UK for some time.

I started this section reflecting on the inadequacy of suitable and sufficient CPD, and you'll notice that it doesn't get the distinction of being identified individually in the KHDA model above, although elements are embedded within it. I'll cover the experiences of teachers when it comes to skills and, most importantly, confidence later in this chapter (see 'Confidence is King'), so I'll avoid duplication for now, except to say that early 2020 experiences quickly sparked the discussion about CPD delivery and frequency and even a fresh discussion about allocating more time during the PGCE or Teach First courses to ensure greater digital skills training is provided before formally entering the profession. There is a lot to be said for that, but given the finite amount of time available to undertake the course, it raises the perennial question, 'At the expense of what else?'

One of the key takeaways throughout the pandemic though, from someone always looking at a glass as half full, was the level of growth mindset that educators in all areas showed. For those unfamiliar with growth mindset (Dweck, 2007), where have you been? I'm just kidding. But to explain, Carol defines two types of mindset – fixed and growth. She explains the drawback of a fixed mindset as follows:

> Your view of yourself can determine everything. If you believe that your qualities are unchangeable – the fixed mindset – you will want to prove yourself correct over and over rather than learning from your mistakes.

What resonated loud and clear, and was amplified across the entire education sector, was the way staff adapted to new solutions as a path to mastery, persisted in the face of daily setbacks, embraced challenges and found support and inspiration from the success of others – all key traits of a strong growth mindset. This mindset has been at the forefront of the willingness to reflect on the innovation potential of EdTech moving forward.

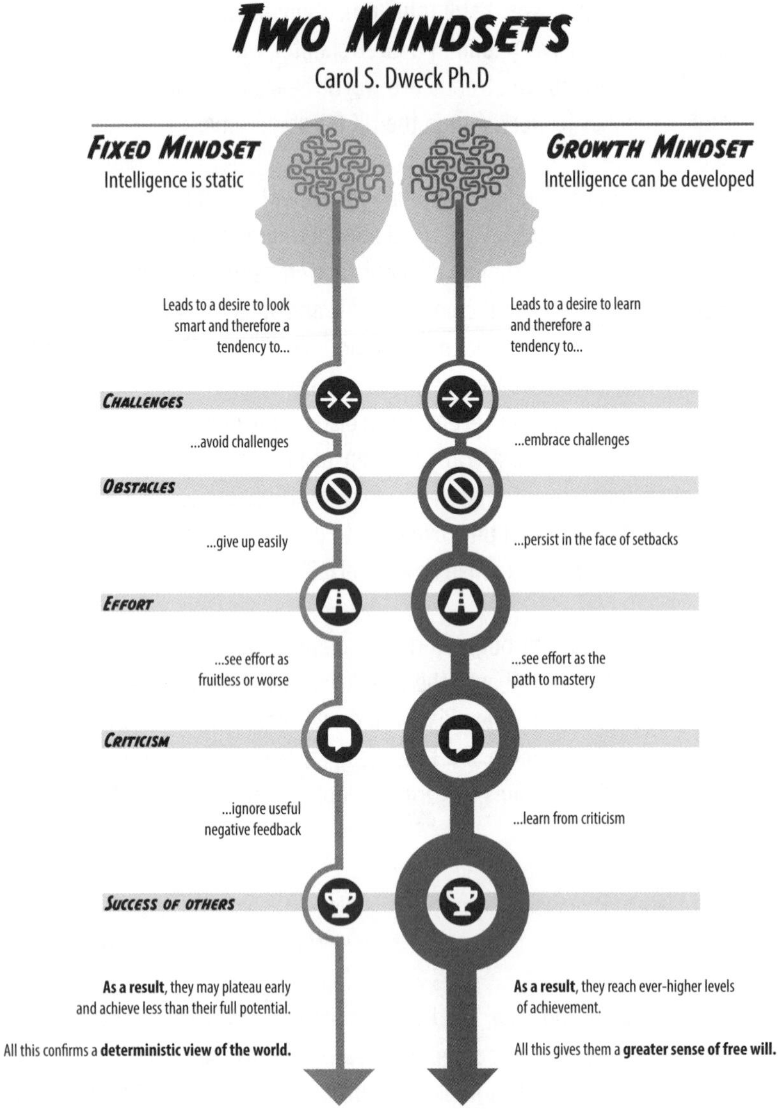
TWO MINDSETS
Carol S. Dweck Ph.D
FIXED MINDSET
Intelligence is static
GROWTH MINDSET
Intelligence can be developed
Leads to a desire to look smart and therefore a tendency to...
Leads to a desire to learn and therefore a tendency to...
CHALLENGES
...avoid challenges
...embrace challenges
OBSTACLES
...give up easily
...persist in the face of setbacks
EFFORT
...see effort as fruitless or worse
...see effort as the path to mastery
CRITICISM
...ignore useful negative feedback
...learn from criticism
SUCCESS OF OTHERS
As a result, they may plateau early and achieve less than their full potential.
As a result, they reach ever-higher levels of achievement.
All this confirms a deterministic view of the world.
All this gives them a greater sense of free will.

One UK initiative that was claimed to have been accessed by over 4000 schools (and indirectly, 200,000 classroom staff) was the EdTech Demonstrator Programme – where schools with experience and capacity to share their digital best practice were funded by the UK government as part of their 'Get help with technology' initiative (www.gov.uk/guidance/get-help-with-technology). Their mandate was to support a range of schools and colleges during the COVID-19 outbreak by helping those who were most in need, had recently adopted an online learning platform or had high numbers of disadvantaged learners. The support package included advice, training, online tutorials, webinars and recorded content. As one well-known retailer puts it, 'every little helps', and I have seen lots of positive comments about the support the Demonstrator schools offered.

By the start of the summer term in 2020, what we knew was that for all the areas of success (and by success, I mean efforts to mitigate, or at least minimise, learning loss during the earlier period), one of the biggest challenges faced by schools was with engagement and the significant number of students who weren't completing any work at home. This was one of the key focus areas for schools to further refine their offers for remote learning. Within the classroom, too, best practice approaches to teaching, whilst maintaining social distancing, were shaped and constrained by physical barriers, movement in the classroom, physical handling of work, resources and so on.

Have I already mentioned that most schools *still* didn't have sufficient kit to support disadvantaged learners? Logistically, sourcing enough devices was always going to be a challenge, but we were late to the party in the UK and the political narrative on delivery success certainly didn't reflect experiences on the ground, which were exacerbated by helpful last-minute reductions in device allocations for each school. As always, schools managed to adapt and over the coming months, promised devices arrived in stages, device donations were received from the business community, local authorities and the broader public, and almost every child who needed technology had access to something.

Subsequent lockdowns presented fresh challenges but without doubt, lessons had been learned and there was improved delivery of online offerings and much greater engagement and evidence of learning. However, one of the key realisations was that the dynamic of a physical classroom with the engagement of a teacher and a room full of children was priceless and could not be fully replicated in an online experience.

Feedback was crucial in mitigating some of the loss from the classroom and the use of solutions that could deliver timely feedback, ideally written or verbal, made a big difference. It was never going to be as instantaneous as in the classroom, but learning doesn't just happen between the hours of 8.45 a.m. and 3.30 p.m. so some of the tools that are now fully embedded have the potential to benefit homework and revision sessions outside of school hours and during the holidays as well.

The education inspection framework's 'outstanding descriptors for effective feedback' (www.bit.ly/3on8aDp) are as follows: 'Teachers provide pupils with incisive feedback, in line with the school's assessment policy, about what pupils can do to improve their knowledge, understanding and skills. The pupils use this feedback effectively' (Ofsted, 2019). Some of the core platforms embedded by schools during the COVID-19 pandemic provided notification of assignments submitted and the opportunity for feedback quickly, and in a variety of formats.

Microsoft's integration of the Forms app within the broader Teams ecosystem, for example, allowed for quick formative assessments with built-in analytics branching alongside to allow teachers to quickly evaluate results, and assignments within Teams provided rubrics alongside automatic scoring and an easy feedback mechanism. Of course, the same functionality was available for Google users and many other platforms, and the unprecedented and necessary demand for these systems during the pandemic resulted in a dramatic acceleration of the adaptation of these tools to become more suitable and valuable to schools.

In January 2021, **Tom Sherrington** (@teacherhead) wrote a great summary of 'crowd-sourced' ideas for checking students' writing and providing feedback (www.TeacherHead.com). Below are the collective ideas he condensed:

Key idea	Description
Live writing in individual shared documents or digital notebooks	Each student has a document they write in online that the teacher can also see and comment on. They respond to tasks in the document.
Live writing in group documents with individual pages or areas	Similar to the above but each student is essentially writing into the same document for the whole class – i.e. a section within a document with their name on it. Everyone can see other students' writing, or they can work collaboratively.
Using Google Slides/ PowerPoint as a writing space	As above but using Google Slides or PowerPoint for writing – each slide pre-prepared with textboxes and students write on their own page. Easy to link to tasks, resources, create templates.
Using forms for short answers	Using forms in Google or Teams to set multiple questions. Students respond to questions individually and the teacher exports into a spreadsheet to see answers from each student.
Using shared spreadsheets	As above but going straight into the spreadsheet – students write short answers into their allocated column (harder to mask students' answers from each other).
Digital whiteboards	Lots of apps/platforms offer this popular feature. Students write on their digital board and then share their responses as requested. Teacher sees them on screen all at once – answers can be long or short; feedback given, verbal or written, to individuals or the class.
Digital sketch pads	Similar to whiteboards but students have their own space on a bulletin board that is easy for teachers to see (all at once or by scrolling through) writing progress in real time and add comments for the whole group or individuals as they work.
Voice/audio notes	Using voice recording add-ons to record verbal feedback – much quicker to record than it is to write the same amount. Students can playback live or later if done asynchronously.
Photographs of work	The basic idea of taking pictures of handwritten work or other non-digital work and then uploading to share with the teacher. Various dedicated apps and use of phones as scanners. Teachers can annotate and return or use for verbal feedback.
Using platform chat function	Making use of the chat function – in Google Meet, Teams, Zoom etc. – to see students' answers to questions. Good for spontaneous, dynamic responses in addition to verbal responses during live sessions. Good tip to use a whiteboard-style countdown so students submit answers simultaneously – so they can't just copy answers given.

Key idea	Description
Standard email	Simply sending work back and forth via email – straightforward for longer pieces such as essays or chunks of work completed offline – not for live lessons.
Question response add-ons	Platforms that allow questions to be set in a dynamic way during live lessons or planned in advance – e.g. multiple choice or written answers. Good for diagnostic questions. Various apps track each student's response, as well as creating poll graphs, word cloud responses etc.
Verbal feedback in live sessions	The obvious method of selecting only certain students or small groups to be on camera at any point to discuss work directly – work perhaps shared via emailed photographs or live document sharing.
Instant surveys for feedback	Using suitable apps to gain instant feedback and evidence of understanding 'on the fly'.

As an aside I would definitely recommend taking a visit to Tom's website – it is jammed with an abundance of resources.

Returning to our reflections during the pandemic period, I read a great article in *Tes* by **Mark Enser** (2021). The article really encapsulated what most of us felt during the period, but he also wrapped up his thoughts by saying: 'Over the 17 years I have been teaching, there has been a constant nattering narrative that technology is going to transform the profession. The past 12 months have shown that, while access to various bits of technology is useful, its utility is primarily in letting us teach how we would otherwise have taught.' Although I thought he was spot on (thinking back to my baseline of technology underpinning teaching and learning), I also couldn't help thinking it was an analysis based upon what seemed a very direct like-for-like swap out. If we think about some of the other tools utilised by schools – solutions like **Century AI** (www.Century.tech) and their personalised learning platform – I don't think it was simple substitution; it was more about augmentation. I base that measure on the fact that even if we were assessing its value back in the physical classroom, it would still be a worthy addition.

Solutions like Teams, Zoom and Google Meet had their limitations but still added value, saved teachers time and facilitated many of the suggestions from the crowd-sourced feedback listed above. New solutions, like **classroom.cloud** (www.

classroom.cloud), were co-produced with schools and joined the mix, adding the extra dimension of not only being able to communicate online with students but also being able to see all their device screens at the same time, no matter where they were. In addition, it gave teachers the opportunity to provide more instant feedback during a synchronous lesson. Technology adapted accordingly and the lesson learned cannot simply be that now we are back in school, these skills become redundant. I suspect Mark didn't mean it quite as I read it, and he shared some great insights, so perhaps I misinterpreted it intentionally to expand the debate. So, here are some questions for you to reflect on. Do you think recording exemplars and building libraries of digital revision resources would be a good step forward? Did you notice some of your learners became more engaged with online quizzing and challenge-based apps? Did some of the new tools in use encourage greater parental engagement than we had ever seen before? Will our parents' evenings return to being solely face-to-face? The list goes on.

It's very easy to validate our views through confirmation bias from within our own schools, but perhaps that isn't always reflective of the broader landscape.

One thing we do know is that most schools now have more devices within their IT estate, and with that comes a greater need for IT tools to centrally manage and maintain them and ensure they are available and suitably resourced for students. It would be a terribly lost opportunity if this rare windfall of extra tech wasn't fully maximised. I probably should reiterate again that we still need to ensure that, no matter what the landscape is, we always keep the focus on evidence to back up the impact of any technology we use.

As we look upwards to further and higher education, the feedback from those utilising learning technology echoes the message of its key role, both past and future. The 2020 survey 'Trends in Learning Technology', conducted by the Association for Learning Technology (www.alt.ac.uk) and published in February 2021, reflected some clear indicators of the direction of travel: 94% of their members indicated collaborative tools would be key moving forward; 85% identified virtual classroom software with blended learning as the key area of growth, alongside media production (72%) and lecture capture tools (67%). For me, the strongest indicator was not just the percentages recorded but the clear evidence of these trends continuing on the ground, as devices have become more available and confidence levels have slowly risen amongst educators.

Confidence is King

There is, of course, a direct link between confidence and levels of productivity and excellence. To mitigate any potential lack of confidence, at the very least we can reflect and be willing to try – when we consider the journey of educators (particularly during 2020 and 2021), their adaptability and willingness to try new technology helped mask some of the inevitable confidence challenges. Schools and educators did an amazing job of adapting their working environment to help maintain engagement with their children, no matter where they physically were, and it would be easy and convenient to simply conclude 'We did it' and leave it there. 'We', of course, includes teaching assistants (TAs), without whose support the journey to this point would have been virtually impossible. (To put that into context, in a report by UCL Institute for Education (www.ucl.ac.uk/ioe, April 2021), 88% of TAs supported vulnerable and key workers' children in school during lockdown, with 51% managing a whole class or bubble on their own to free up teacher time for prep and delivery of remote learning. Enough said.)

In practice, what we learned is that confidence-building is always going to be difficult without the right continual professional development. The sudden focus on identifying, implementing and then actually utilising digital technology during the pandemic was not only a very steep learning curve for many just getting to grips with the plethora of new apps and systems: it also shone a spotlight on the obvious disconnect between the training that was needed and what had been provided beforehand.

Not for the first time, what we had to recognise was that when implementing change within our schools, especially where it involves new solutions and technologies, we must also budget for and invest in quality training (and allow adequate time for that training) – not just consider the technology in isolation. Incidentally, by training, I don't mean a simple one-off inset (in-service training) session.

Imagine the concept of a Premier League football team operating like educators – they would expect (and receive) a day of training at the start of the season, covering a range of new skills and techniques. They would then be expected to play competitive matches (lessons) every day of the week, and be equipped to put those new skills into practice perfectly for every game for the rest of the season. In practice, they know regular training is the best way to fine-tune and build confidence in skills.

As **Angela Watson** (@Angela_Watson) articulated on her 'Truth for Teachers' podcast: 'Confidence comes through building capability, and capability comes through repetition.' She hits the nail on the head (and has some other great discussions on her podcast – www.thecornerstoneforteachers.com). Repetition is key and, pre-pandemic, many hadn't used the apps and tools we had to learn to adapt to, or if they had, only minimally. I want to emphasise the point that even when new skills are acquired, they need to be put into practice regularly – 'If you don't use it, you lose it' – or the training is wasted. 12–18 months post-pandemic, I wonder how many teachers will continue to confidently embed some of the solutions we discovered.

Some of the concepts or approaches shared here I've been advocating for years and (courtesy of this book being a catalyst to expand my own reading and research) some are new ideas I've added to the mix. When it comes to seeing or hearing new ideas, I'm a self-confessed sponge and a committed lifelong learner.

At this point, I'd like to reflect on the work of Mandinach and Cline and how that aligns with teacher confidence in the use of technology. In 1996, they published 'Classroom Dynamics: The Impact of a Technology Based Curriculum Innovation on Teaching and Learning', where they identified that technology based professional development programmes become successful when they are focused on the teacher's stage of use, help to inform and change teacher behaviour and are field-based. Mandinach described four stages of technology use: survival, mastery, impact and innovation.

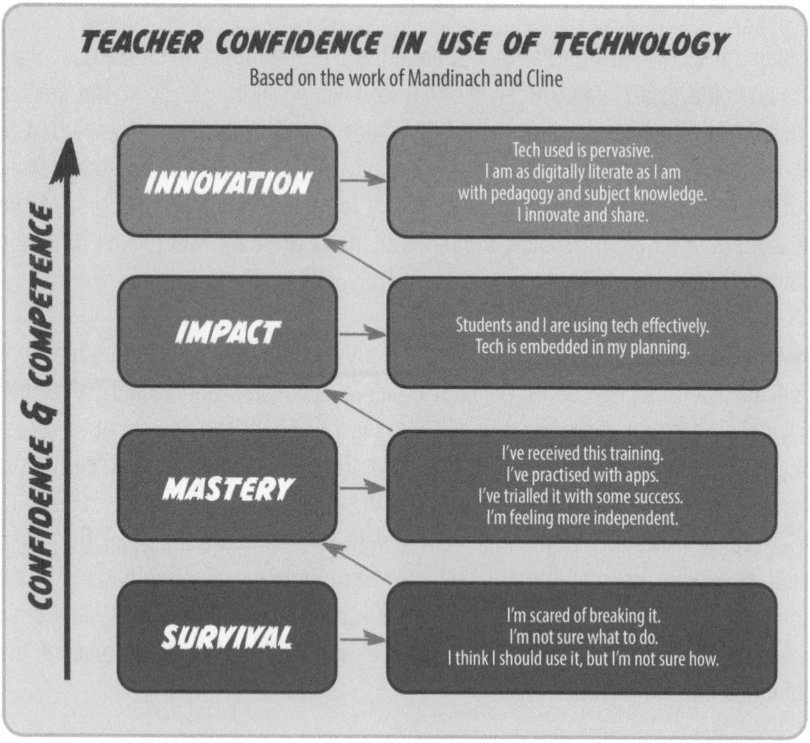

- **Survival** – A teacher in the survival stage struggles against technology; is overwhelmed by problems (everything that can go wrong will go wrong); doesn't change the status quo in the classroom; uses technology only for directed instruction; has management problems planning how 30 students will access devices; has unrealistic expectations, believing that technology use by itself will result in higher academic performance.
- **Mastery** – A teacher in the mastery stage has increased tolerance to technical problems; begins to use new forms of interaction with students and classroom practices; has increased technical competence and can troubleshoot simple problems.

- **Impact** – A teacher in the impact stage regularly incorporates new working relationships and classroom structures; balances instruction and construction; is rarely threatened by technology; regularly creates technology-enhanced instructional units.
- **Innovation** – A teacher in the innovation stage modifies his or her classroom environment to take full advantage of technology-enhanced curriculum and learning activities.

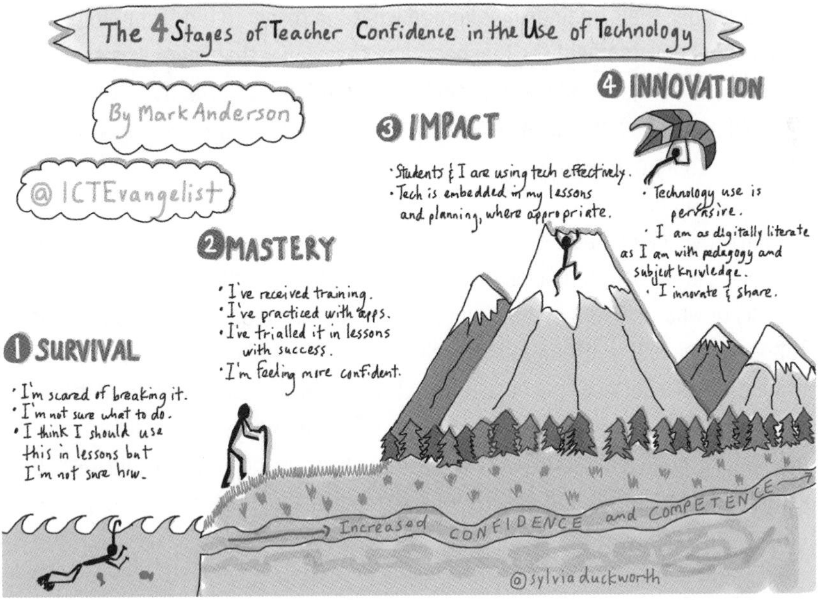

I can highlight this 'journey' with an image by **Sylvia Duckworth** (www.sylviaduckworth.com) (based on work by **Mark Anderson** – www.ictevangelist.com) who produces the most engaging sketch notes and images on all aspects of educational life. Sylvia very kindly gave me permission to use her image in this book and it visually describes the above journey perfectly.

So, what are the salient lessons we have learned and, more importantly, what have schools done to help address this challenge? Certainly, taking stock of current skills and knowledge before attempting to move forward (by undertaking a staff digital skills survey) proved to be a key part of the process – identifying existing skills and levels of confidence and where CPD and support can best be provided to build on those skills.

As part of my research and reflections on what seemed an obvious point, I reviewed numerous studies, including one in 2007 entitled: 'Professional Development in Integrating Technology Into Teaching and Learning: Knowns, Unknowns, and Ways to Pursue Better Questions and Answers' by (Lawless and Pellegrino). Unsurprisingly, their findings highlighted what the research (on technology-based professional development for teachers) had revealed – that there was still a long way to go in terms of educating teachers in methods of effective practice with technology, particularly with regard to the identifiable impacts these activities had on teaching and learning.

One narrative which has been gathering momentum recently concerns extending teachers' digital competencies prior to qualification. A study called 'A multilevel analysis of what matters in the training of pre-service teachers' ICT competencies' (Tondeur et al., 2018) attempted to evaluate the potential impact of greater digital training for pre-service teachers (PSTs). The aim of the study was to identify the correlation between *perceived* ICT competencies among PSTs, taking into account their background characteristics (age and gender), their ICT profile (e.g. their attitudes towards ICT) and the various strategies they experienced during their teacher training by:

- Using teacher educators as role models
- Reflecting on the role of technology in education
- Learning how to use technology by design
- Collaboration with peers
- Scaffolding authentic technology experiences
- Continuous feedback

Based on a survey of 931 final-year PSTs in Flanders, the multi-level analyses indicated a positive association between the strategies listed above and ICT competency; the more PSTs were exposed to the strategies during their teacher training, the higher their *perceived competence* in utilising ICT for learning processes and strengthening their instructional practice.

Gender and age did not affect ICT competence in educational practice. In addition, the results revealed a positive impact on PSTs' attitudes towards ICT in education, including ease of use and their ICT competence in educational practice. This does reinforce the view that there is a need to elevate this narrative into the broader discussion of teacher skills for the 21st century.

Putting longer-term strategies to one side for a moment and focusing on the here and now for educators, there are six barriers (often described as inhibitors) to teachers using technology in class:

- **Conviction** – Teachers need to genuinely believe that the addition of technology in the classroom will add value, enhance their lessons and, therefore, be beneficial to teachers and students alike.
- **Time** – No surprises here – teachers often feel they simply don't have the time to learn the skills to use new technology.
- **Accessibility** – Some of this has been mitigated by the recent influx of additional devices, but in some cases, there still isn't sufficient access to devices or connectivity to be able to plan their use on a regular basis.
- **Training** – Another unsurprising barrier – teachers need access to sufficient training with the technology they intend to use.
- **Support** – Teachers need access to technical and (sometimes) peer-based support to build skills.
- **Confidence** – As already mentioned, practice and frequency of use will build confidence once the other barriers are addressed.

Although none of the above barriers are a surprise, often when challenges are summarised in this way, they are easier to tackle and track over time.

John Tomsett and Jonny Uttley's book 'Putting Staff First' is a very worthwhile read. The authors highlight the importance of developing a professional learning programme, one strand of which might well be for digital skills. They quote

Tom Bentley's 1996 speech at the National College of Teaching and Leadership Conference where he explained: 'Once you have found your core purpose, change your school's existing structures to accommodate your core purpose rather than contort your core purpose to fit around your existing structures. If the development of teaching and learning is your priority, then you have to find the hours during the school week for your staff to work on their practice. No ifs, no buts – with a will, any logistical challenge can be overcome.'

I agree with this view completely and it simply must be done, although as professionals we should also take a degree of ownership for our own professional development. In addition to the formalised considerations above, I've summarised a few top tips from my personal experience and the suggestions of many others:

- **Less is more** – focus on small steps on a few key solutions. It's better to master a few than attempt to upskill on too many too soon.
- **Use your PLN** – join Twitter and follow peers who regularly share advice and tips – #EdChat #EdTech etc.
- **Learn online** – there's an abundance of brilliant online discussions, shows and podcasts that share best practice. I have listed a few top 10s at the back of this book.
- **Attend conferences** – for networking and to gain greater exposure to solutions and good quality CPD.
- **Flag bearers** – find the key users for each solution – those who are confident in their use and actually use them – as a go-to point for guidance.
- **Don't be intimidated** – if others appear to learn faster. Stick to what works for you – you will get there eventually.
- **Giving is receiving** – share experiences and tips with colleagues in the same situation as you and they will reciprocate. You'll also pick up some great ideas.

I cannot stress enough that if you don't give people the time and resources to build confidence using EdTech, you are never likely to see the true impact it can have – which is ironic in education, when CPD and development of pedagogy normally have excellent frameworks for support.

It's also crucial that we don't just focus on solutions and approaches we already know about; we must also ensure schools remove the silo mentality (the tendency for departments within an organisation to not share information or knowledge with other departments, groups or colleagues) and provide a conduit to share awareness, insights and guidance on any other tools or solutions that might assist others.

As an example, the **JISC 'Teaching staff digital experience'** insights survey of further and higher education teaching staff in the UK (JISC, 2020), showed 84% of respondents had never used nor had familiarity with any of the key assistive technologies, e.g. screen readers, dictation, screen magnifiers or any alternative input devices. In our drive for accessibility of digital solutions, this is just one small example of why it's so important that educators know what's available in the broader toolkit to support their learners.

Asking the Right Questions

With the caveat that no two schools are alike, if you're thinking about undertaking a staff digital skills survey, then asking the right questions is obviously key, especially if you repeat the process over time and track the changes in feedback and confidence. A proactive board of trustees might ask for evidence of those changes too.

With that in mind, I've assembled a selection of example questions that focus on the successful delivery of remote education. You can adapt these depending on the current challenges or insights you want to gather. Use a quantitative rating of e.g. 1 to 5 on relevant questions (1 being lowest, 5 being highest).

- Do you have high-speed internet at home?
- Do you have access to a device for working at home or delivering online lessons?
- Have you been able to access the resources you need from school?
- Are you confident with the technology you are using for online teaching?
- Are you confident with all the applications/software you use each day?
- How helpful has the school been in offering you the resources to teach from home?
- How helpful have your co-workers been while teaching from home?
- How stressful do you find teaching remotely?
- How stressful do your students find learning remotely?
- How was your experience teaching students from home compared with teaching at school?
- Have you accessed any online courses or CPD for best practice online?
- How well were you able to maintain a work-life balance while teaching remotely?
- Do you enjoy teaching your students remotely?
- What level of engagement have you achieved from your students online?
- How practical/suitable is the environment at home while you are teaching?
- How important to you is the role of technology in remote learning?
- How important to you is face-to-face communication while teaching remotely?
- How often do you have a one-to-one discussion with your students?
- How helpful have parents been while supporting their children's remote learning?
- What specific training do you feel you would benefit from?
- How else can the school support you further?

Hopefully, the focus and emphasis of the survey questions post-COVID-19 will be more about confidence with the tools utilised within the classroom, but ensuring the skills acquired during the pandemic aren't just forgotten is also key to future flexibility – even if that's just selectively using the elements which worked well for things like online exam revision classes in the holidays or retaining the online parents' evening software.

Keeping everyone safe

Schools nowadays have so much more responsibility than simply educating our children. Pupils' social, emotional and mental health (SEMH) is (rightfully so) a high priority, as is recognising the inevitable balance of opportunities and risks that the digital world presents. Before I go any further here, there is no denying that the most important tools for keeping our children safe are the eyes and ears of both parents and professionals.

I've tried to break this section down into a few key areas that we now know are the building blocks to keeping children safe: our moral and legal obligations; the availability of constantly evolving technology to support those obligations; the practical approaches schools can adopt; the critical task of equipping our children as digital citizens; and online resources relevant to this key subject.

It goes without saying that anyone involved in education supports the view that every child matters and does everything they can to provide a safe environment where every child can thrive. As with everything in education, we have legislation and policies that shape the expectations of our sector. When it comes to digital and online safety, the key guidance and regulations covered here include **KCSIE** (Keeping Children Safe in Education) and the **Prevent duty** in the UK and the **CIPA** (Children's Internet Protection Act, 2000) in the US.

Keeping Children Safe in Education (KCSIE) – UK

Usually updated annually (at the time of writing, the latest version was January 2021), this covers statutory guidance for schools and colleges on safeguarding, safer recruitment, handling allegations and the role of the designated safeguarding lead. Of particular relevance in the context of this book is annex C, which covers online safety (Department for Education, 2021, p. 102).

Now I know that everyone working in English schools will have been required to read KCSIE annually, but on the basis that it is possible (fingers firmly crossed) readers may be from other parts of the world, I am including the salient points for completeness:

> The use of technology has become a significant component of many safeguarding issues. Child sexual exploitation; radicalisation; sexual predation: technology often provides the platform that facilitates arm. An effective approach to online safety empowers a school or college to protect and educate the whole school or college community in their use of technology and establishes mechanisms to identify, intervene in and escalate any incident where appropriate.
>
> The breadth of issues classified within online safety is considerable, but can be categorised into three areas of risk:
>
> - **content** – being exposed to illegal, inappropriate or harmful material; for example pornography, fake news, racist or radical and extremist views;
> - **contact** – being subjected to harmful online interaction with other users; for example commercial advertising as well as adults posing as children or young adults; and
> - **conduct** – personal online behaviour that increases the likelihood of, or causes, harm; for example making, sending and receiving explicit images, or online bullying.
>
> Governing bodies and proprietors should be doing all that they reasonably can to limit children's exposure to the above risks from the school's or college's IT system. As part of this process, governing bodies and proprietors should ensure their school or college has appropriate filters and monitoring systems in place.
>
> Whilst considering their responsibility to safeguard and promote the welfare of children and provide them with a safe environment in which to learn, governing bodies and proprietors should consider the age range of their pupils, the number of pupils, how often they access the IT system and the proportionality of costs vs risks.

The appropriateness of any filters and monitoring systems are a matter for individual schools and colleges and will be informed, in part, by the risk assessment required by the Prevent Duty.

Prevent duty – UK

The Prevent duty guidance (Home Office, 2021) sits under the Counter-Terrorism and Security Act 2015. Released by the UK government in March 2015, the guidance was made a legal requirement on the 1st of July 2015. It requires schools (and other specified bodies) to put measures in place to 'prevent people from being drawn into terrorism'.

Within the guidance, there are four key areas where schools need to demonstrate appropriate action:

- **Staff training** – The designated safeguarding lead needs to undertake WRAP (Workshop to Raise Awareness of Prevent) training to be able to provide advice and support to other members of staff.
- **IT policies** – The need for school leaders to ensure that children are safe from terrorist and extremist material when accessing the internet in schools.
- **Working in partnership** – Working with local safeguarding children boards (LSCBs) who can provide advice and support to schools on implementing the duty.
- **Risk assessment** – The need to assess the risk of/identify children in danger of being drawn into terrorism.

CIPA (Children's Internet Protection Act, 2000) – US

In 2000, Congress passed the **Children's Internet Protection Act** to help combat the growing concern about children accessing obscene or harmful content on the Internet. Any K–12 school or library which accepts certain federal funds is required to use and evidence their 'technology protection measure' on every computer connected to the Internet, to block or filter any harmful content, before they can receive E-rate funding.

Under CIPA, schools must ensure their internet safety policies include monitoring the online activities of minors and educating children about appropriate online behaviour, including cyberbullying awareness and interacting with other individuals on social networking websites and in chat rooms.

- Schools are also required to implement suitable measures within their internet safety policies that address or mitigate:
- Access by minors to inappropriate matter on the internet.
- The safety and security of minors when using electronic mail, chat rooms and other forms of direct electronic communications.
- Unauthorised access, including hacking, and other unlawful activities by minors online.
- Unauthorised disclosure, use, and dissemination of personal information regarding minors.
- Measures restricting minors' access to materials harmful to them.

If we consolidate the above, from both sides of the Atlantic, we can quickly identify the common strands: an expectation and obligation to provide – *where appropriate* – suitable filtering and monitoring to avoid access to inappropriate content; a mechanism that warns if a child is accessing information that may place them at harm; key training on the risk for staff and awareness for children as part of the curriculum delivery.

Staff safeguarding training is already well embedded in schools and I don't think I can add any additional value or insight here, other than to share some useful resources that may support any ongoing CPD later in the book.

In terms of appropriate filtering and monitoring, this is a particular area of focus, experience and interest for me, and I have spent the last 15 years fine-tuning software solutions to meet this need. It's also important to recognise this is another area of sensitivity for many, with the words 'snooping' and 'spying' often part of the discussion. We will cover the two approaches (filtering and monitoring) to meet statutory obligations and then clarify what this looks like in reality.

Filter or Monitor

Filtering and monitoring are often used as two interchangeable words to describe the same activity, but that really isn't the case. Filtering is designed to restrict or control the content a user can access on the internet. It works by preventing pre-determined words, phrases and URLs from being delivered to the user. Normally this is a device or service working at the perimeter of the school network – imagine it as a sieve over the school's broadband connection as data enters or leaves the building.

Monitoring, though, rather than blocking data at perimeter level, operates in the background, typically at a local level (per device). Unlike filtering at the perimeter, which is just viewing web activity, monitoring solutions look out for pre-set words and phrases across all user activity – internet, email, Word, etc. – and because it works at a local level, it can alert a member of staff if anything unusual appears.

Filtering and monitoring both rely on a master database – a list of inappropriate websites to block (filtering) and keywords and phrases covering everything from bullying, radicalisation, suicide, self-harm, grooming, gambling, racism, drugs etc. to raise an alert (monitoring). Obviously, new terms and websites appear all the time, often regionally, so the master database needs to be updated regularly.

Most dedicated school ISPs (internet service providers) already include a layer of filtering which ensures many harmful websites are blocked. In the UK at least, most are members of organisations like the **Internet Watch Foundation** (www.iwf.org.uk) who provide lists, updated daily, of sites that contain, in their case, any evidence or risks surrounding child sexual exploitation.

It's probably important at this juncture to explain how these tools work and dispel a few myths. Imagine everything you type scrolling across the screen in a single, continuous line. With monitoring software running in the background, every word is screened and processed in the device's virtual memory. Once the words disappear off the end of the screen, they are lost forever – no history of what was typed is retained. But if the monitoring software detects a word or phrase that matches with its database of key terms – for example, 'suicide' – the software will capture that specific keyword (and a few words either side, to provide context) and trigger an alert with all those words included. The alert then goes to the designated safeguarding lead (DSL) or nominated staff member(s) who can then manually review and decide whether it poses a concern or is considered a false positive.

There are quite a few well-established solutions that offer this functionality, including **NetSupport DNA** (www.NetSupportDNA.com), but I have always felt that operating at this level alone offers far more likelihood for false positives than for useful warnings. For example, 'You da bomb' and 'build a bomb' would both trigger the same alert – obviously, context is key. So, we took it a stage further and came up with a model based around 'contextual intelligence' – in essence, it considers a *range* of variables to identify whether there is a real risk or not.

Imagine two scenarios: Billy opens a PowerPoint slide during a lesson and adds a title: 'The History of the First Nuclear Bomb'; meanwhile, Freddy, a vulnerable child, opens a selection of websites on an unsupervised PC in the library after school and researches several sites on how to make a bomb. I have exaggerated the two examples for effect, but hopefully what it highlights is that the risk posed to both children includes a number of variables:

- What was the word or phrase typed?
- Have they searched for similar terms previously?
- Was it on a supervised or unsupervised device?
- Was it inside or outside school hours?
- What software application or website was it triggered on?
- Was it triggered by a vulnerable child?

Using contextual intelligence, we would hopefully conclude that although it might be the same keyword that was triggered, a one-off trigger by a child on a local application during a lesson with no previous related triggers would rate as a much lower risk than a child repeatedly researching the same topic out of school hours in an unsupervised setting online. Solutions like NetSupport DNA allow you to focus on triggers that need immediate intervention – such as clear self-harm, suicide risks etc. – and get a sense of other underlying topics that you can intervene on as required.

As an important aside, when looking for a monitoring solution for your school, also consider protecting your EAL (English as an additional language) students by ensuring that whatever software you choose includes languages within its master database which match your cohort and demography.

As mentioned at the beginning of this section, the most powerful safeguarding tools in a school are the eyes and ears of teachers, and all the above contextual intelligence is what an observant teacher would process within seconds. So, let's be clear – no matter what the tool, intervention and action alerts will always require review by a suitably qualified professional to validate their relevance.

Whilst I have hopefully clarified the distinction between filtering and monitoring, it is also important to provide a conduit for children to report their own concerns. With the growing prevalence of online bullying, for example, having a system in place for any child to report concerns to a trusted professional at school is fundamental to empowering our children. Think about the narrative and profile of events in March 2021, when the **'Everyone's invited'** website (www.everyonesinvited.uk) was created and over 11,000 children posted their experiences of sexual abuse and harassment at school. Many of the allegations on the website referred to sexual harassment carried out against young women by young men at their school or university. Alongside the need for government intervention and review, providing tools that facilitate a safe voice for children is vital, now more than ever. I'm in a very fortunate position to be able to influence product design and functionality to help alleviate risk, so our core safeguarding solution also features a 'Report a concern' option.

As I mentioned earlier, there are a broad range of considerations and, when appropriate, educational technology has a big role to play. But alongside this, there are two other areas to consider – promoting a culture and focus on safeguarding within your school and encouraging and facilitating good digital citizenship.

You can tell a lot about a school by what happens when you walk through the door – how visitors are welcomed and processed, what information/leaflets are shared, the visibility of signage etc. – and some of that good practice has had to be extended (and tailored) to fit the virtual world too. For example, the school's website, key safeguarding and contact information and how easy it is to find policies and anything else relevant. Schools have necessarily had to adapt how they share key information with students – for example, at the start of the school day or online lessons, they present reminders about policies and best practice – and are now providing better resources to parents and carers.

EdTech can play a helpful role here – whether it's utilising your digital signage around the school, sharing key messages, or providing access to appropriate and relevant safeguarding information on every digital desktop. We are all conscious of the need to share sensitive supporting information on topics like FGM (female genital mutilation) in places where a child is likely to feel more comfortable accessing it, and it makes sense to ensure that the same applies within a virtual space.

The last core topic in this section is digital citizenship (#DigCit) – the responsible and safe use of technology. It's a broad topic that covers security, privacy, conduct and an understanding of the risks that exist. I usually try to frame this discussion around the idea that if every child was taught to be aware of what they should/shouldn't share online, had the skills to question the validity of the information they read, knew how to conduct themselves and treat others online and be aware of the risks that exist – the need to discuss many of the topics earlier in this chapter would largely disappear.

#DigCit

The **International Society for Technology in Education** (www.ISTE.org) outlines nine elements of digital citizenship to help students navigate online:

- **Digital access** – Advocating for equal digital rights and access is where digital citizenship starts.
- **Digital etiquette** – Rules and policies aren't enough – we need to *teach* everyone about appropriate conduct online.
- **Digital commerce** – As students make more purchases online, they must understand how to be effective consumers in a digital economy.
- **Digital rights and responsibilities** – Students must understand their basic digital rights to privacy and freedom of speech.
- **Digital literacy** – This involves more than being able to use tools. Digital literacy is about how to find, evaluate and cite digital materials.
- **Digital law** – It's critical that students understand how to properly use and share each other's digital property.
- **Digital communication** – With so many communication options available, students need to learn how to choose the right tools according to their audience and message.
- **Digital health and wellness** – One important aspect of living in a digital world is knowing when to unplug. Students need to make informed decisions about how to prioritise their time and activities online and off.
- **Digital safety and security** – Digital citizens need to know how to safeguard their information by controlling privacy settings.

It's not a one-way street either. The benefits of digital citizenship for children extend far beyond the child. When we help children develop healthy practices online, we're also creating a better space for everyone else they might interact with. Alongside that, because technology is so prevalent in schools, the onus often falls on teachers to prevent many of the online risks. Teaching digital citizenship is the most effective way to have a positive impact, reduce inappropriate behaviour and keep our children safe online.

I'd like to give a shout out here to **Henry and Danielle Platten** and the team at **GoBubble** (www.GoBubble.school) who promote a safer and healthier digital community for children, where positive actions and activities are rewarded and children can develop their digital skills within a safe and controlled environment. It's a brilliant example of educational technology being used in a positive way to develop and promote good digital citizenship and I highly recommend it.

Vicki Davis (www.coolcatteacher.com) wrote a great summary for **Edutopia** (www.edutopia.org) in 2017 where she summarised teaching her children the **'9 key Ps'** of digital citizenship, namely:

- Passwords
- Private information
- Personal information
- Photographs
- Property
- Permission
- Protection
- Professionalism
- Personal brand

This really encapsulates all the strands well and provides a good outline of the knowledge and experience children need to act responsibly online. I've included a link to the article in my reference section. Separate to Vicki's points, I've included a nice visual summary below encapsulating some of the key messaging for students.

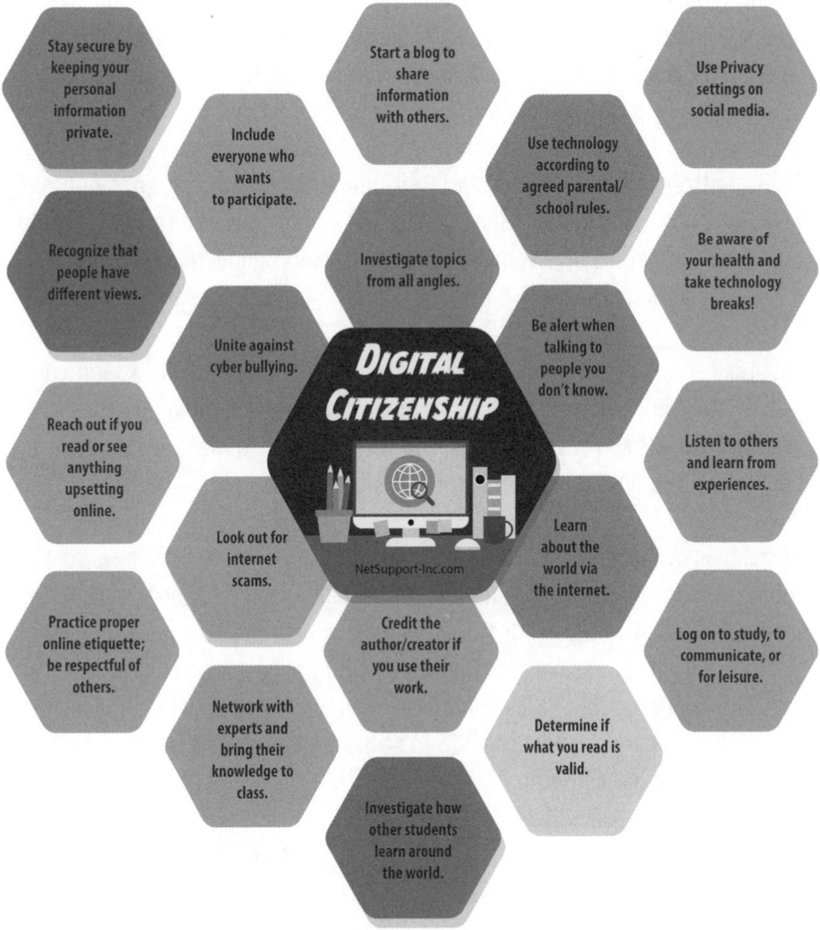

Developing digital leaders within schools is also another positive way to embed the right culture and mindset. Within the context of age ranges, students selected as digital leaders can help to embed the use of technology across the school. They attend regular meetings, support other pupils, sometimes teach members of staff, run assemblies and other whole-school events, and lead on improvements in eSafety provision. Digital leaders can not only support their peers with the use of technology: they can also be an active and positive voice on all aspects of eSafety, present reports to the school governors, perhaps evaluate new software or devices, or act as e-ambassadors. I am sure you get the idea that developing digital leaders is an absolute win-win and not enough schools recognise the benefits.

'Digital citizenship has become a large umbrella term which encompasses a variety of definitions. Historically speaking, the digital citizenship conversation begins and sometimes ends around online safety followed by a long list of don'ts.

Just take a look at the language used in your school's tech policies and you'll immediately know if your school community approaches digital citizenship reactively or proactively. When working with school communities, we start by changing those **don't** statements into **do** statements.

By changing the narrative this way, our focus shifts towards how to use technology for good by learning how to become a force for good online. Our language around safety needs to extend beyond creating safe spaces online by additional layers to the conversation like digital wellness, and how to prioritize my time online, to how to evaluate the accuracy, perspective and validity of digital media and social posts, to how to be more inclusive and embrace multiple viewpoints and engage with others online with respect and empathy.'

Marialice B.F.X. Curran, PhD
Founder and Executive Director, Digital Citizenship Institute
www.digcitinstitute.com

Safe places

Although this book is a journey through what we have collectively learned over the years and where we now need to focus, I'd like to highlight a few useful (and trusted) sources that I think are worthy of being shared with staff and, where appropriate, with students.

Grooming	www.childnet.com www.nationalchildrensalliance.com www.nspcc.org.uk
Cyberbullying	www.nationalbullyinghelpline.co.uk www.stopbullying.gov www.bullying.co.uk www.americanspcc.org
Sexting	www.internetmatters.org ('Look at me' report) www.nationwidechildrens.org
Social media	www.rsph.org.uk www.childmind.org www.connectsafely.org

As a conclusion to this topic, I also wanted to share an article I wrote for SecEd Magazine (www.sec-ed.co.uk) in August 2020, which focused on creating safe ecosystems around any provision of remote learning.

SECED MAGAZINE, AUGUST 2020

Remote learning:
Creating safe digital ecosystems

Remote and online learning is likely to be with us for some time as the COVID-19 fight continues. As such, we must continue to prioritise eSafety. Al Kingsley asks how we can promote and deliver safer digital ecosystems in schools and better develop pupils' digital citizenship.

The near future looks likely to be one of blended learning approaches, in which online learning will continue to play a key role as pupils return slowly to the classroom.

It may be a paradox, but the first thing to be clear on with creating safer school digital ecosystems is that technology is not the key. It is actually the approach and behaviour that makes the difference when supporting safeguarding in digital situations with remote learners. Some of the best ways schools can do this would be by continuing to promote a positive digital culture as they interact with students, observing any changes in behaviour and engaging as much as possible.

In the current climate, teachers have had to step outside of their comfort zones very quickly and work out how to make things safer and more usable as they go along. Sharing what we have learned during this time – perhaps collaborating with peers, either locally or online – is something to be encouraged as we all navigate this new learning landscape. Here are some things to consider.

Soft skills at the forefront

A safe digital ecosystem is really about manners and ethos. Start with a whole-school approach. Get buy-in from everywhere: parents, teachers and staff. Make sure that your senior leadership team is on board and that there is time for training to address the variables and build teachers' confidence in using both the tools and the accompanying offline approaches.

Communication is at the heart of the process, and from the senior leadership team all the way down the chain, stakeholders need to know that they can contribute to the conversation and that their voice will be heard.

Central to all of this is the need to remove finger pointing and blame when things do not go to plan and look upon it as a learning experience from which you will all come out stronger and wiser. If children are placing themselves at risk, then it is a collective responsibility and it is all about supporting them with the best strategy we can as we prepare them for the world ahead.

Developing teachers' online confidence

For teachers who are not particularly confident with technology to start with, heaping the responsibility for online safety on top of that can be pretty daunting. However, it can really help to remove the notion that the teacher should be the one who knows everything, and instead recognise that the whole school community is learning together and needs to accept constructive critique to move forward.

I would offer two tips for developing confidence online. First, learn from the students themselves. Despite their only real focus (and motivation) being gaming, YouTube and social media, they can often give a real steer and insight into some of the things they encounter, which teachers can both learn from and adapt.

Second, make use of online platforms such as Twitter. There is an awful lot of noise there of late, but it is by far the best platform to connect with education professionals. The key is to be selective and choose a dozen or so educators to follow, from whom you can discover best practice and ideas that you can modify to suit your own context.

Remote online safety

When learners are separated from us at home – perhaps during a local lockdown or for those who may need to isolate – the need to be alert to their online safety is even more pressing. Teachers will naturally take ownership and responsibility when seeing any kind of inappropriate online behaviour – and sometimes that will mean talking to parents too.

Do what you would normally do: engage with the student first, then the parents, if you can – and, if needed, seek advice from team members. Once you know the details, you can help the student to reflect on the consequences of anything they have done online and the potential danger they have exposed themselves (or others) to.

Being at home in isolation can give all of us – not just students – a false sense of security and protection when online. Sometimes, it is easy to forget the extent of the audience it ties into and there is no real sense of comeback. Extending the reach of remote online safety as part of your digital ecosystem is all about reinforcing the messages of digital citizenship with your students, and even repeating the basics – 'Don't put anything online you wouldn't want your grandparents to see!' – ensures they are at the forefront of students' minds as they go online independently.

Technology, again, is not the key to encouraging remote online safety; it is the behavioural approach that makes a difference. Schools eSafety advisor Alan Mackenzie talks about 'online disinhibition' which links behaviour and psychology. Essentially, this represents a lack of judgement or a restraint when communicating online. Sometimes, children use that disinhibition in a benign way: they are more comfortable sharing personal feelings or emotions when remote to someone. However, sometimes it can be more toxic, whether taking the form of inappropriate comments or behaviour online or failing to assess the risky elements of what they have been doing.

So, it is clear that we need to ensure that we are focusing on the digital citizenship element of embracing a safer ecosystem: focusing on young people's awareness of the implications of their online activity and making sure that, in terms of our safeguarding and oversight, we treat students the same way as we would in the classroom.

Empowering parents

One of the phrases that is often referred to when it comes to looking at the broader digital ecosystem and eSafety is 'getting comfortable with being uncomfortable'.

Getting parents on board to support online safety at home can be challenging, but empowering them with knowledge such as how to apply filters or parental controls – whether with written guides, online training using Microsoft Teams or other tools – can really build their confidence.

Reminding parents of simple strategies for home online safety – e.g. the benefits of children using digital devices in a communal space (so they can keep an eye on their activity), setting expectations when doing remote learning and having home-school agreements – is really valuable, not just for the student, but for their own sanity too!

Digital considerations

In a purely digital sense, schools still of course need to think about things like online filters and context-based keyword monitoring. Good tools should also allow a platform for children to proactively report their concerns and reach out for help when they need it.

It is also important for schools to consider what new applications they have implemented, especially in the context of the personal data being stored (in line with GDPR) and the extent of the protection offered for that data.

In the haste to deliver solutions to meet remote learning needs within a safe digital ecosystem, leaders should not forget the essential preparations (particularly communication and training) to ensure all staff are confident in the tools and approaches they are or will be using.

And remembering that the centre of it is not based on anything digital, but on collaboration, sharing, engagement and knowledge-building, will help schools establish best practice for flexible – and safe – remote teaching and learning in the future.

Planning technology and infrastructure

While I was busy doodling my spider chart on what to include in this book, it's probably no surprise (given my background) that this topic was one I needed to get off my chest. Unfortunately, I'm old enough to remember the days of network installations that were nothing more than a daisy chain of all the devices with a 'T-piece' on the network card at the back of each PC, and they were all linearly linked together. This was the backbone of early networks like **ARCnet** (www.arcnet.cc) and 10BASE2 ethernet. It was very simple, but one break in the chain and the entire network went down.

Single Point of Failure

That concept stays with me whenever I think about the lessons learned and the planning required for a school's infrastructure. We often add all kinds of front-line resources, devices and apps without considering the foundations of our school's infrastructure and how they might flex or fail. I'm sure any IT administrators reading this book will identify with my comments.

If we unpick this on a very simple level, a school/MAT/district's IT infrastructure isn't that complicated – it's managing the devices connected to it that creates the real overhead. Much of the 'simple' stuff is invisible to non-IT technicians and, as a result, considerations such as hardware/software refresh cycles (the frequency with which they need to be replaced), IT capacity, compatibility and connectivity rarely hit the headlines.

I've been involved in new free school builds where you really do start with a blank canvas so to distil it down into AI speak (reflecting on topics without the distraction of the minutiae and detail), if you want connectivity in your school, there are a few interconnected elements.

Let's start with a desktop PC with a network card. All that's required is a cable from the network card to a socket on the wall (which could support 100Mbps or 1000Mbps data rates, so you need regular or fast sockets). Each of those sockets throughout the school is wired via a physical cable to a junction box, called a 'switch', which acts as the main connection to the school's network and distributes data to all the connected devices. If you had a mature network and ran out of sockets to plug your PCs into, you might also have a 'hub', which acts like a dumb switch that also broadcasts frames of data to all ports – think of it like an extension block providing extra sockets.

So, PC to socket, socket to switch. All the switches across the school building are cabled physically back to a central 'patch panel' (which is just a connection point for all the cables) and then feed into bigger switches which connect all the areas of your school network together. That's it – that's what we call a 'network', or 'LAN' (local area network). If you plug all your PCs in and then plug a 'server' (a computer that serves information to other computers) into another socket, you then have data sharing. If you have two schools with LANs and want to connect them together, you have different types of switches called 'routers' which allow you to connect multiple LANs together, thereby creating a 'WAN' (wide area network). If you want access to the internet, you then add a 'firewall' to the connection between the internal network and the outside world so you can protect inbound and outbound traffic and control who has access to resources on your network.

When schools decided that tablets and laptops were the way forward, networks had to adapt and provide WiFi capability. WiFi works by having 'access points' (APs) throughout the school, each having a 'range' that allows them to be seen by a mobile device. Just as before, the access points are wired into a socket or a switch and are then connected to the main network. WiFi has moved fast over the years and the type of WiFi you have will impact the speed of connection as well as how many devices each access point can support simultaneously.

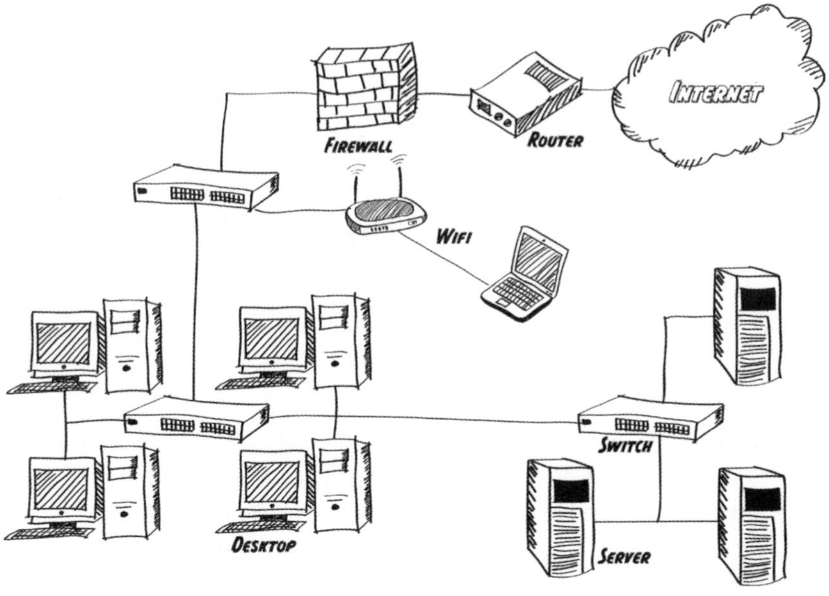

So why the networking lesson? Because every time a school decides to purchase new iPads or Chromebooks, the type of access points used, their capacity, the number deployed for connectivity, the speed of the switches, the capacity of the network – all these elements need to be factored in to ensure the IT infrastructure can deliver what it should. Too often, schools spend their budget on new devices and applications without considering the vital cost of ensuring the network can handle the additional capacity and load – and when it can't and the device user has a poor experience, the device is (wrongly) blamed.

Just like the refresh cycle required for our end-point devices (laptops, tablets, PCs), schools have become much more aware that the technology behind the scenes has advanced significantly too and budgeting to maintain and upgrade the infrastructure is equally important.

The basic concept of networking and infrastructure is relatively simple (most IT administrators will agree that it's not the tech but the users that present the challenge). In truth, the real wizardry and workload is in managing and maintaining all the users, policies, access rights and processes that make a network usable, granular and secure.

With the connectivity challenges considered, the next part of the evolutionary debate is about where we store our precious data.

To Cloud or Not to Cloud

That is the question...and it's become more of a discussion point in recent years. I could be direct and tell you there isn't a choice, but as with most things, there are some choices available.

What is the cloud? In essence, it's simply a term used to describe servers that host applications elsewhere than at your physical location. When I say you don't have a choice, what I'm really saying is that so many services and facilities are now in the cloud that every school *will* be using the cloud to some degree. Where there is choice is over where you store your core operational data and what type of applications you use. I cover some of the narrative about this within the 'From a Vendor's Perspective' chapter, but essentially, you can either host your servers locally within your school or have them hosted in the cloud by someone else. Although hosting on site potentially gives you greater control and security over your data, cloud hosting avoids the need to purchase servers, maintain them or have any associated business continuity considerations (e.g. flood or fire damage).

Operationally – e.g. when dealing with accounts, HR, student information, IT management etc. – you have the choice between using local or cloud-based systems, but for other aspects (such as curriculum content or personalised learning solutions), they are typically only available if you head to the sky.

The COVID-19 era has highlighted the benefits of making key resources accessible from anywhere, not least when staff and students are off site. It's not a simple black-and-white consideration as, technically, using tools to make your physical server on site accessible to staff at home is, in effect, creating a private cloud. Microsoft and Google have made a compelling case for utilising their solutions in the cloud, including the central management of key resources and easy-to-access file storage (OneDrive, Google Drive etc.).

Things get trickier when you start thinking of all the curriculum resources schools sign up to, some of which are downloaded, some accessed online. With online resources, student (and staff) data is shared within those solutions to use them effectively. This is where policies and procedures need to come into play. Schools should undertake a data protection impact assessment (DPIA) or equivalent to consider what data is being shared, where it is stored, how long it will be there and, of course, who has access to it. It's a risk-based view, but in our haste to meet online needs during the pandemic, it's probably fair to say this process became a bit more casual for some schools. Data security and privacy are extremely important considerations and having a tight grip on what data is held, where it is held and who can access it should be integral to all decisions made.

It cannot be stressed enough that there should be a renewed focus on security across school IT systems, not least as they increasingly seem to be a target for hackers. Make sure school policies are appropriate and well-communicated – it is critical to enforce rules on password complexity or (on key systems) multi-factor authentication (the requirement to further validate a user's identity beyond just entering a username and password), anti-virus and anti-malware solutions, and other integral security measures. Larger MATs should consider penetration testing – using a third party to test broader network security – as this can also pay real dividends. One good program to consider for either self or external validation of security approaches is **Cyber Essentials**(www.ncsc.gov.uk/cyberessentials/overview),a UK government-backed scheme for best practice with information security.

Like it or not, the landscape has dramatically and permanently changed and before adding any kind of educational technology to your school, it's important to have the discussion about what it might mean to the IT infrastructure – where it will be located, whether it creates a security risk and (one consideration that is often forgotten) whether it will create any CPD requirements for staff. As a reminder, CPD shouldn't just be for the teachers – IT staff need to be given the opportunity to gain product training so they are able to fully and effectively provide support.

No matter what approach a school or MAT takes in terms of IT infrastructure, I have always advocated that knowledge is power and it's crucial to have a full view of all the devices across the entire IT estate – what they are, where they are, how often they are used, what software is installed on them, what data is stored on them – and with that knowledge, plan appropriate refresh cycles and effective deployment.

In 2003, we developed our first solution to provide oversight of an IT infrastructure – **NetSupport DNA** (www.NetSupportDNA) – and it has evolved over the years to include the full gamut of features needed. It always surprises me how many schools consider buying more devices without knowing how effectively their current devices are being used. Knowing which PCs are rarely used and could be deployed elsewhere can prevent unnecessary IT spend. Knowing the specification of devices and whether they can accommodate a memory expansion helps with planning and, just as importantly, knowing which devices haven't been used for months (or, conversely, which have been left on all night for weeks) can help with security and potential energy savings.

Choosing the right solutions and applying the best IT strategies should, of course, be research- or peer-informed and for extra validation, I'd always recommend two brilliant groups for any network managers where peers can provide tips and advice – the **Association of Network Managers in Education** (www.ANME.co.uk) and **EduGeek** (www.Edugeek.net).

We still need to think about how recent events have shaped our approach to the deployment of technology, even if we are now comfortable that it's all working well and is suitably supported. Let's take the example of classrooms in my trust. Some years ago, we chose to replace individual laptops for staff with front-of-class PCs in some of our schools. That decision saved money and worked well at the time, but of course when COVID-19 hit, some staff were without a work device to use at home. Although that situation couldn't have been predicted, this one example links to a couple of strands – teaching confidence and wellbeing.

Confidence is relatively easy to articulate – walking into a classroom with your own device set up and ready to teach means confidence will be higher than it would be having to rely on a memory stick and hoping the technology will be compatible and work as expected. Familiarity with technology shouldn't be understated. Additionally, a desktop PC wired into a smart board doesn't provide the opportunity for teacher mobility around the classroom, or screen beaming, or access to other amazing app ecosystems. It would be far more effective to provide a point of connection where different devices could be used depending on topic and content.

The same goes for wellbeing, in terms of the hours of preparation involved in teaching. It's not difficult to appreciate that without access to the latest technology, the preparation process is going to be much slower and less fluid if staff are forced to rely on personal tech in the evenings. Making the workload as manageable as possible has been another lesson learned and factors into decision making now. The choice of which device, and its connectivity within the classroom, needs to be so much more than just a technical consideration (I'll expand on this in more detail in my chapter 'Planning Ahead').

Last, but not least, as I mentioned security in this section, here is a list of the worst passwords used in 2020, courtesy of Alan Mackenzie's *Ditto* magazine:

1234567890	12345678	123456
12345	password	000000
123123	picture1	Iloveyou!
111111	123456789	

Some Money Saving Tips

Budget will always be a sticking point for schools wishing to realise ambitious EdTech plans. The **British Educational Suppliers Association** (www.besa.org.uk), in their report entitled 'ICT in UK Maintained Schools', stated that when it comes to ICT, the biggest challenge for secondary schools is securing funds. The magazine Education Technology (August 2019) identified that 67% of schools felt that securing funds was the single biggest challenge they faced in trying to move forward (up from 56% in 2018). Unsurprisingly, second on their list (at 49% of respondents) was adequate training for teachers.

Soon after the BESA report was published, I wrote an article for Education Technology (www.edtechnology.co.uk), presenting half a dozen tips for smarter EdTech. I recognised and highlighted that school leaders were under increasing pressure to do more and spend less. While well-managed deployments of technology can help to achieve this, there are numerous ways schools can make sure they are squeezing every drop of value from their investment in EdTech.

The key advice was to forget budgets – focus on strategy. While there are practical tips and tools that can be used to minimise the costs of running school technology, the only real way to ensure the greatest return on investment is to treat EdTech as a long-term strategy, rather than a series of short-term wins. Having said that, if you need to drive some of those short-term options, consider the following.

Upcycling unused devices

Before spending, it's always good practice to undertake a thorough audit of your existing IT assets to identify what you have and, more importantly, how effectively they are being used. A decision can then be made about whether those assets could be upgraded or redeployed, rather than replaced. As an example, an IT manager at a primary multi-academy trust recounted one particular primary school selecting ten aging laptops in a storeroom. Instead of replacing them at a cost of circa £3000, they upgraded them with solid-state drives at a fraction of that amount. They were subsequently put to good use in an intervention room where they are used to aid children's research.

Reducing software costs

Software is intrinsic to the day-to-day running of all schools. Some solutions will remain essential, but others, perhaps curriculum resources in a particular department, will be superseded or cease to be used on a regular basis and need to be changed. Does your school have visibility of which software licences are being renewed automatically, despite being unused? Keeping track of your installed licences can avoid unnecessary license renewal, helping to reduce expenditure. There are several tools available to help manage this process, like NetSupport DNA as mentioned earlier, which can be considered as part of the ongoing IT strategy.

Reducing time costs

Supporting IT teams in proactively maintaining a school's network can also save money by avoiding extended periods of downtime caused by IT issues. There are IT management tools which can proactively alert when any changes occur on specific devices or the network. Being aware of these minor issues gives IT staff an opportunity to be proactive and prevent such small problems from escalating into larger ones, thereby minimising wasted downtime for teachers, students and support staff – and the precious time saved can then be used more productively (and economically).

Reducing wasted energy costs

Did you know that a standard desktop PC left on continuously, day and night, consumes more than 500kWh per year? Based on just ten PCs, this equates to over £700 (at the time of writing), so it makes sense to use energy monitoring tools to identify how many PCs are left on out of hours – thereby reducing costs. Power management policies can then be implemented to power off selected PCs automatically at the end of the day, and turn them back on (all at once, or in stages) the following morning. In addition, 'inactivity policies' – creating rules for systems to sleep, log out or power down if they have been inactive, factoring in term time dates to ensure devices aren't left on over the holidays – will help deliver further savings. Naturally, as the device landscape changes towards a more portable footprint, it's likely that this will become less of a factor.

Reducing print costs

In an average size secondary school, the cost of student printing alone is estimated at around £100k–£150k per year. Unnecessary printing can often spiral out of control, incurring high paper and toner costs. Using print monitoring software, a school can see exactly where costs are being incurred and make necessary adjustments, such as reducing the number of printers available for student use, setting print limits, preventing duplicates being printed, and so on. The advent of larger MFPs (multi-function printers) with more granular controls, alongside a move towards digital resources, has helped ease some of the pressures on this aspect.

Although I appreciate these are all small cost-saving suggestions and probably ones already being considered, it's important to recognise that often the largest cost savings are a result of lots of marginal gains and rarely the result of one specific intervention.

Improving communication

'Communication let me down, and I'm left here' said education experts Spandau Ballet in 1983, and let's be honest, there are few organisations who could claim to have a perfect system and approach. Ironically, it's harder than you'd think to identify and list all the opportunities for communication, not least because it's an ever-moving landscape and every year new mediums join the mix.

I am not intending my EdTech diary to suddenly morph into a self-help guide on effective communication strategies in schools, either in an interpersonal or management context. Rather, my intention is to reflect on how technology can foster and facilitate greater levels of communication.

We can probably divide communication flow into three main areas:

- Staff to staff
- Staff to student
- School to parent (and community)

As **Leah Davies** eloquently shared in an article on effective communication (www.bit.ly/3yneCPx):

> Being able to communicate is vital to being an effective educator. Communication not only conveys information, but it encourages effort, modifies attitudes, and stimulates thinking. Without it, stereotypes develop, messages become distorted, and learning is stifled.

Well put and helpful in considering how much importance and value is placed on the same concepts in terms of the above categories – can we use EdTech, for example, in a positive way to modify parental perceptions, attitudes and engagement with the school? I think so!

In terms of the various methods of communication typically used within a school, although there may be more, I would summarise the key platforms (and not all digital) as:

- School planners
- Emails
- Telephone calls
- Face-to-face meetings
- Social media pages
- Digital newsletters
- Report cards
- School websites
- Parent portals
- School prospectuses
- Student learning journals (primary)
- Mobile apps

As I've said before, EdTech can be the communication facilitator, but it still relies on the human element for the actual message and content. As we used to say in the early days of computer programming, 'If you put garbage in, you'll get garbage out.' That still holds true (except perhaps in the case of Auto-Tune in the music industry...).

It's good practice to have a communication policy because it not only encourages us to segment our stakeholders (parents, students, staff, governors, etc.) based on their specific needs and requirements but also helps us to identify the most appropriate methods of communication for each group.

Things to think about:

- How often do you receive communications from each of the stakeholder groups?
- What methods do you currently use to communicate with parents or other staff?
- Do you use different methods for internal versus external communication?
- Are you responding to enquiries outside of your core working hours?
- Are communications having a negative effect on teaching workload?

Staff to staff

I mentioned silo mentality earlier (the tendency for departments within an organisation to not share information or knowledge with other departments). It's a non-physical barrier that often exists between staff and departments within a school.

In the last decade, the opportunity to share data and resources and become connected has increased more than ever, particularly in terms of shared storage of resources (e.g. SharePoint servers, OneDrive, Google Drive, Dropbox etc.). However, what has also become prevalent is this silo approach, where individual departments follow a different path based on their specific needs rather than within the context of the whole school. It works but of course over time the opportunity to share broader resources across the school becomes more difficult, disjointed and, from a data protection perspective, IT teams then need to manage and monitor multiple platforms. Joined up thinking creates communication cohesion, reduces disparity between different departments and therefore reduces operational overhead. Even so, it doesn't have the biggest impact on communication.

Much more significant is the dialogue between colleagues – the discussions about projects, individual students or life in general – which is restricted when work pressures dictate a revolving pattern of home-car-classroom-car-home. During the period of COVID-19 lockdowns and restrictions, it was surprising how many staff found that the introduction of, for example, Microsoft Teams provided an unexpected and accessible way to re-open lines of communication. During a process of discovery and increasing familiarity with the platform, and growing confidence using it, much more dialogue took place between staff in and out of school hours. It also became a catalyst for non-work-related dialogue that helped maintain relationships, supported wellbeing and boosted morale. In a typical weekly working schedule, with so many pressures and responsibilities, solutions like Microsoft Teams and Google Meet proved effective in lowering the silo walls and facilitating better staff communications.

During a time of extreme and rapid change caused by the pandemic, clear and effective top-down communication from the SLT was key to reassure and support the workforce. Confidence comes from knowing what's happening at any given time. Email (particularly late in the day or, even worse, in the evenings) can often have a negative effect on staff wellbeing, and it can also be unnecessarily difficult keeping track of an email dialogue which creates a lot of inbox traffic. Consolidated platforms provide a much easier way of communicating that also enables instant collaboration, with the added benefit of audio and video options. Although I'm stating the obvious, these platforms mitigated much of the geographical divide for staff when working from home.

The same applied to the continuity of local governing body and trust board meetings. With the need to avoid face-to-face meetings due to the risks associated with them, key discussions and strategic oversight could continue online – and delivered a hidden benefit: attendees got home much more quickly afterwards, because they were already there!

Some schools use WhatsApp groups (or equivalent) for maintaining contact and social dialogue with peers, but probably the biggest change recently has been the explosion of PLNs on platforms like Twitter (www.twitter.com).

Hashtag

For peer support, advice, best practice, sharing resources and general mental wellbeing, being part of an online community has huge benefits, and, of course, aligning a school's presence on the platform can be a great way of showcasing the school's values. I can't begin to express how much I have learned from so many willing peers by participating in online communities, and I'd encourage you to engage if you haven't done so already. As with all forms of social media, it's good to know how to find the right topics and people. #Hashtags are the key to Twitter. #EdTech is an easy one, or #EdChat for a broad connect, but check out **TE@CH with ICT** (www.teachwithict.com) too – Simon shared a great article with 50+ popular hashtags for educators (see 'References' for the link at the end of the book).

It's not just about Twitter either. In the US particularly, good old **TikTok** (www.TikTok.com) has a rapidly growing educator community, with loads of short instructional videos and classroom ideas.

Alongside Twitter and its new live feature, a range of apps have appeared to facilitate live teacher community discussions, and a particularly interesting one, **ClubHouse** (www.JoinClubHouse.com), is worth checking out. It's a new type of social network based on voice – where people around the world come together to talk, listen and learn from each other in real time.

When it's an Emergency

There are times during the school year when we need to plan and test our ability to communicate fast (unfortunately, this isn't usually to announce good news or showcase student work) and to ensure we have a system for rapid communication when there is a potential risk or threat to students and staff. Although this is a rare occurrence, it forms part of our safeguarding obligations.

In 2017, the UK teaching union NASUWT called for all schools to have a clear protocol to be used in the event of a serious risk occurring on or near school premises, and since then, more and more schools have adopted a 'lockdown' plan. This falls under the broader term of 'invacuation', a process for getting everybody to somewhere safe and secure within a building. Typically, children and staff will be able to stay in their classroom during an invacuation and the school day can continue in some shape or form.

The lockdown aspect takes things a step further and requires everyone on premise – staff, children and any visitors – to be moved away from the potential danger to a place where they can't be seen from outside the building. I must confess, when I think of this topic I immediately imagine terror threats or unprompted attacks, but more typical scenarios for an invacuation might be:

- Nearby air pollution or similar due to a fire or chemical release.
- A dangerous animal (usually a dog but I'm happy to include a lion) in the grounds.
- An incident or civil disturbance in the area that might affect the school.

Lockdown might be implemented if there is:

- A disturbed or intoxicated person trying to get into the school (this could be a parent or a stranger).
- An intruder on the site.
- An internal threat from a student.

I'm not going to explain how schools respond to the lockdown alerts in terms of windows and doors and movement of staff and children as that's readily available online, but in our broader discussion on technology and communication, I've highlighted how EdTech can and does play a role in supporting this process.

Many schools will simply raise audible alerts via, for example, a fire alarm using a different tone, which can be a practical way of sending a whole-school message to stay put and move away from the windows. However, for larger sites or where movement to a secure place is needed, a granular approach can be more effective. Many schools and districts now employ the use of desktop alerting software to instantly communicate a message to all registered devices. A desktop message or alert takes screen focus and remains visible until the user has either acknowledged it or it has expired (a fixed duration can be specified). On-screen alerts can also include school display screens (covered further down). Alerts can be pre-configured to quickly trigger if, for example, someone in reception is confronted by an individual – a simple triple key press to send an automated warning to every connected device.

Desktop alerting software should be factored into all emergency protocol considerations. They are typically low cost, and a good solution will also be able to flex for more regular non-emergency use: 'The mail server is down – we are aware and we are working on it'; 'Mr Smith, there is an urgent call for you in the office'; 'The fire alarm test will activate at 10 a.m. – there is no need to respond'; etc. Desktop alerting software can be used as an integral part of daily school life to improve communications where immediacy is key.

Additionally, schools can use the same alerting tools in a pre-configured group in situations such as when a newly qualified teacher (NQT) is struggling with behaviour in a classroom – they can send a message to key staff who can then do a walk-by or step in to assist.

There are quite a few solutions available, like **NetSupport Notify** (www.NetSupportNotify.com) and for US readers, I'd also recommend the **National School Safety Centre** (www.schoolsafety.us), who provide a range of free resources and guidance.

Sharing the love

Staff to student

Communication within schools is often positive (the opportunity to share student success for example) and also informative (such as the school dinner menu, health and safety guidance or upcoming school events). EdTech plays a huge role in this, replacing the humble noticeboard with interactive, rich and engaging displays. In my opinion, digital displays contribute significantly to the fabric of a school and I am a vocal advocate.

There are typically two main questions/barriers raised: cost and maintenance.

There is a perception that you need to purchase expensive systems to deliver effective digital signage and there is a natural concern that maintaining those systems and keeping them up to date with content requires capacity and resources most schools simply don't have.

In terms of cost, there are some good solutions available right now, from something as simple as an Amazon Fire TV Stick plugged into an LCD screen (less than £40/$55), to a Chromebox, the inbuilt WebOS within a smart TV and of course regular Windows or Mac devices. In some schools, to save spending anything, old LCD monitors have been repurposed and face outwards through classroom windows so parents waiting for pickups can see the latest news and updates. Reception areas and staffrooms reflect school life and there is a constant flow of highlights to share, either happening within the school or from its social media channels. Anyway, the point is, you don't need to spend a lot of money upgrading the messaging around your school.

In terms of ongoing maintenance and resources, there are solutions that are simple to use and allow teachers to easily post content directly from their classroom to supplement the main school messaging. It can be as easy as an app on an iPad to post photos or success stories and is something that should be embraced. Solutions like **Trilby TV** (www.TrilbyTv.co.uk) in the UK or **YoDeck** (www.YoDeck.com) in the US are a good starting point.

As discussed earlier, platforms like Microsoft Teams, Google Classroom, Zoom, classroom.cloud etc. have all played a significant part in enhancing communication flows whilst breaking the restrictive tether of location. They have not only enabled schools to develop frameworks for good pedagogical communication flows and feedback loops but also created significant opportunities for peer interaction and made a positive impact on social, emotional and mental health as a result.

School to parent (and community)

We have learned a great deal about the benefits of good outward communication from our schools. Arguably, the COVID-19 pandemic was a timely catalyst to focus our efforts on trying new approaches and improving that information flow.

In sharing my thoughts on this, I am conscious of making sweeping generalisations and appearing ageist (especially as I have plenty of grey hair myself), so I'll aim for realistic diplomacy. Based on career progression and accumulated skills, the most senior staff within any organisation 'tend' to be older and, therefore, more experienced.

Social media, YouTube and live chats are all relatively new and something our children and youngest staff use constantly. The discussion surrounding communication (and marketing) strategies for schools, MATs or districts is often led by those less invested in (or familiar with) current trends for communication. However, they are also generally the ones making the decisions about what they think is best for the school. On one level, that's fine – someone has to decide. But if you're not an active Twitter user and haven't gained any value from it, or an active member of community Facebook groups etc., then the odds are you won't see the benefit of the school investing in any of those mediums. Added to that, when it comes to Ofsted inspection time, the focus tends to be on website content and all the right policies and links being accessible, rather than on other mediums.

We also need to be mindful of the age demographic of many of our primary school parents and the likelihood that they will be using an app on their smartphone to keep in touch and up to date. The pandemic was a useful (if unwelcome) catalyst to revisit this subject again and think about what our parents now use to communicate and where our community groups are. During the pandemic, there was a noticeable (and necessary) increase in school-to-parent communications via social media and community groups, sharing information on school activities, fundraising or simply keeping the parent-school engagement going.

None of the above removed the value of bulk emails/announcements to parents for key events, and tools like **ParentMail** (www.parentmail.co.uk), **Parentapps** (www.parentapps.co.uk), **Schoolcomms** (www.schoolcomms.com), **Groupcall Messenger** (www.groupcall.com) or **SchoolStatus** (www.schoolstatus.com), to name but a few, continued to be indispensable. But in response to the challenge of conducting parents' evenings remotely, another variant of tools suddenly came into play.

The pandemic provided an opportunity to set a new standard for parent-teacher meetings, where every parent would hopefully be engaged, and no one missed out. For many families, traditional face-to-face meetings can be a source of tension. Balancing busy schedules and childcare can result in many failing to attend; for separated families, meetings can prove awkward or difficult, particularly for the child. Delivered remotely, however, many of these issues cease to exist – parents don't need to book childcare, arrange travel or even engage with other families/ family members, and the latter point proved to be a popular win for many parents.

Time is always a major factor – for staff, parents' evenings can overrun and prove a stressful and tiring experience; for parents, waiting around, watching appointments overrun or believing others are jumping the queue can be frustrating. Remote parents' evenings are far more structured: everyone gets a time slot, and when that time is up, the meeting concludes automatically. It removes a lot of the stresses associated with the process and, as a result, it improves the quality of the conversation, allowing teachers to share their resources and the session to flow more naturally without any concerns about being overheard.

The necessary change to remote parents' evenings seems to have been regarded as a huge success by parents and staff and I anticipate that in future we might see a blended model, with perhaps two online and one face-to-face version each year.

In terms of tools, unsurprisingly a few new solutions came to market in 2020 and whilst I can't recommend one over another, popular solutions included **Parents' Evening Manager** (www.iris.co.uk), **SchoolBooking** (www.schoolbooking.com), and **SchoolCloud Parents Evening** (www.parentseveningsystem.co.uk). I suspect they are all here to stay.

Before I conclude this section, I'd encourage you to check out an insightful report written by **Alison Kington and Agato Mleczko**: The Advantages of Successful

School-Community Relationships: Findings from the Include-Ed project. Over a four-year period, they undertook research on the impact and outcomes from a range of strategies, including those focused on fostering strong links with the community and community involvement in their schools. Although communication alone is not a measure of a school's vision and values or approach to community engagement, it is an important factor and one that builds an early perception of a school from the outside. The report findings demonstrated that there was a strong connection between participation by the school in the community and a positive impact on existing issues they had identified. A number of these related to transforming the neighbourhood, such as:

- Gaining accessible housing information
- Availability of support for parents and families
- Healthy eating, living and health provision
- Employment opportunities
- Supportive environment
- Social opportunities/outlets
- Social opportunities for specific cultures
- Parental views and opinions

The case study data also showed that there was a correlation between school/community participation and positive transformations related to lifelong learning, empowerment, and social cohesion, such as:

- Engagement in family education courses
- Achievement of qualifications e.g. ESOL, NVQ
- Parents studying at the local university
- Confidence building
- Community participation and independence
- Voluntary involvement in school activities
- Support for specific groups of parents
- Socioeconomic and cultural divisions

Expanding further, participation in family education courses increased parents' confidence, especially for those from certain cultural backgrounds. Both parents and pupils stated that involvement in school activities had increased their confidence in dealing with staff and school-related issues. It also had a positive impact on the number of parents willing to attend school-based clubs and school trips to support and enrich the curriculum. In addition, participation significantly improved communication processes, not only through the activities organised but also because of the interaction generated by this. Thus, the school had become a meeting place for information exchange, bonding members of the community together.

That has got to be a good thing in my book.

Promoting wellbeing and SEMH

This is the final section in this chapter and some of the topics already covered feed positively and negatively into the wellbeing discussion. According to the Oxford English dictionary, wellbeing is 'the state of being comfortable, healthy or happy', which feels too narrow and 'in the moment for me'. It should perhaps be broader and include a sense of purpose, happiness with and level of control over one's life.

In the guide '**Measuring Wellbeing: a guide for practitioners**' by the New Economics Foundation (2012), it is stated that 'wellbeing can be understood as how people feel and how they function, both on a personal and a social level, and how they evaluate their lives as a whole'.

> Challenges are what make life interesting. Overcoming them is what makes life meaningful.
>
> JOSHUA J. MARINE

Although it may seem a less than obvious topic to include, EdTech has been proven to have both negative and positive influences on wellbeing. Emails being sent/received out of core hours from school leaders (or the DfE) and the instant accessibility that Teams and other solutions provide aren't always such positive things. The rise of technology and smart devices at our fingertips has made us significantly more accessible, and consequently, a healthy work-life balance is increasingly more difficult to achieve.

Policies and best practice on communication protocol – how and when mediums are used and the expectations of recipients – need to be implemented, managed and balanced so that messages about wellbeing aren't simply a tick in the box or a token gesture. In addition, the communication mediums used by schools should proactively provide secure and safe places (channels/groups) for staff to interact and support each other on a social basis too, so that personal networks can be maintained. This has proven a real positive in many schools and should be embedded for the long term.

Employee assistance programmes which provide external support to staff should be clearly communicated and as accessible as possible. The transformation and changes with EdTech should be rolled out with full support in terms of training and time for staff to learn, adjust and feel confident and competent. In other words, before you give someone something extra to do, first consider what you plan to take away from them to keep the balance.

The research paper '**How EdTech can support social and emotional learning at school and at home**' published in the International Journal of the Whole Child (Caukin et al., 2020) by **Nancy Caukin, Leslie Trail and Ashlee Hover** identified how EdTech can play a pivotal role in supporting social and emotional learning (SEL) and provide the flexibility of learning at school, home and elsewhere. They shared a variety of free and low-cost solutions for a range of ages that promote the competencies outlined by CASEL (www.casel.org), to help facilitate consistent and ongoing opportunities for students to practice SEL skills. SELwas also covered in the article '**How to Implement Social and Emotional Learning at your School**' (Elias, 2016) by **Maurice Elias** on **Edutopia** (www.Edutopia.org).

I've shared a few apps from the research paper and added extra apps to get you started.

SEL competency	App	Age	Description
Self-awareness: recognising one's emotions, behaviours, and ways of thinking; understanding one's strengths and weaknesses; having a sense of optimism and confidence	Emotions by Avokiddo	4+	Gives young children opportunities to explore feelings and emotional connections through several characters and props; activities explore the cause-and-effect relationship of facial expressions.
	Emotionary by Funny Feelings	4+	A great resource for children with special needs; includes descriptions of emotions and funny feelings; allows users to draw an emotional 'selfie' to show how they are feeling.
Self-management: regulating emotions, thoughts, and behaviours; controlling stress, self-motivating; setting and pursuing goals	Breathe, Think, Do by Sesame Street	4+	Teaches young children about problem-solving, self-control, planning, and time on task; players are presented with different scenarios in which a monster character must regulate his/her emotions using the breathe-think-do technique.
	GoNoodle www.GoNoodle.com	6–12	Short physical activities that provide children with brain breaks to increase concentration and attentiveness; activities require children to cross the midline of their bodies which engages both sides of the brain.
	Stop Breathe & Think by Tools for Peace	11+	Emotional check-ins and personalised recommendations for meditation, sleep, breathing, and yoga.
	See Saw www.SeeSaw.com ReallySchool www.ReallySchool.com	All	Allows classroom teachers to communicate with parents and share student work. It also helps build SES by engaging students in a variety of work styles.

Social awareness: empathising and understanding social and ethical norms for behaviour	Touch and Learn Emotions by Innovative Mobile Apps	4+	Includes photographs that represent four different feelings per page – players match verbal cues with appropriate photos to help them identify body language and understand emotions.
	Forest – Stay Focused by Seekrtech	4+	Productivity app – the main purpose is to teach users to stop using phones as distractions and to be socially present; the more time users spend in real life, the more their virtual tree grows.
Relationship skills: establishing healthy relationships with diverse people; using appropriate communication, cooperation, and negotiation Responsible decision-making: making good choices about social interactions, understanding consequences, and considering the wellbeing of others	Peppy Pals Social Skills by Peppy Pals		Includes games, books, videos; animals socialise, take care of each other, solve problems and explore emotions.
	The Middle School Confidential Series by Electric Eggplant		The books/apps are designed by a teen expert, Annie Fox, and presented in a graphic novel sequence in which readers follow a group of seventh-grade friends as they navigate friendships, families, and school.
	Class Dojo www.classdojo.com		This can be a school-wide or classroom teacher program that helps schools build a culture around helping students monitor their own behaviour. Teachers track the good things that students do through the day and the concerns that they have about student behaviour. Students are able to earn points for good deeds, good work, and good behaviour.

There are many more but hopefully the summary above highlights a few areas where EdTech has the potential to play a positive role. As always, impact is often shaped by implementation and we will cover some of that in my later chapter on 'Planning Ahead'.

From a governance perspective

The past 20 years' experience in school governance at different levels, combined with the challenges of COVID-19 during 2020 and 2021, really brought into focus the additional layers of responsibility and support that educational governance provides.

If you are not familiar with the term, now is a good time to introduce the concept of 'critical friend' – encapsulating the real-world yin and yang of 'challenge and support' when managing the performance and direction the school is taking, alongside its primary responsibility of ensuring the safety and wellbeing of everyone involved with the school. If the last 18 months taught us one thing, it was that the first part of every discussion should always begin with 'How are you?'

If we imagine the school structure as a pyramid, in a standalone school the governing body sits at the top, supporting and challenging the head and senior leadership team. Within the wider context of a multi-academy trust, the pyramid includes: the governing body (or academy committees for each school), focusing on curriculum standards and local issues; the trust board above them, with broader responsibility for the finance, HR and key operations; and members who set the strategic vision for the entire trust at the very top.

Although I'd like to delve into the broader topic of governance – best practice ideas, feedback and much more – this book isn't about governance in detail, only where it plays a role in reflecting on the last two years and how it can support the onward journey in terms of the effective use of technology within our schools. That said, if you would be interested in learning more about governance, or perhaps are already involved in school governance, feel free to visit **www.schooltrustee.blog**, where I regularly share articles and insights covering all aspects of governance. I've maintained the blog for a few years now; it's a useful repository to share different challenges or updates and does seem to have been helpful to others continuing their governance journey.

Within the context of this book, we all recognise that governors are primarily concerned with ensuring schools deliver the best and broadest possible curriculum to feed into our teaching and learning outcomes. Of equal importance (I believe) is the breadth of stimulating and memorable experiences we provide for our children during their school career. It's not just about the acquisition of knowledge – it's as much about equipping our children with the experience and skills they need to flourish in life. When we consider these priorities through the lens of technology, an innovative digital strategy has the potential to provide more than just digital skills for our children: it can also provide many more opportunities to add to and enhance those experiences.

In the same way that our trustees and members have a strategic plan for the whole academy trust, a digital strategy follows a similar path, and I will discuss and expand on this within the chapter 'Planning Ahead'. As this section is more concerned with reflecting on the lessons we have learned, particularly over the last 18 months, I intend to keep the narrative within that context.

In an effort to keep this book as regionally agnostic as possible, although I might mention Ofsted or the Department for Education from a UK perspective, the same layers of oversight and policy feed into educational systems regardless of whether you are an international school in the UAE or a school district in the US. As our journey through COVID-19 progressed, there were mandated expectations – throughout the world – of how schools should use digital technology to support and maintain a level of teaching and learning. There was a global response to try to mitigate the digital divide by accelerating the availability of devices for those young people who had no access to tech. Digital inequity is just one element of an innovative digital strategy and this area, along with a number of others we will explore, will formulate a target list of questions you should be considering for discussion at your board or governance meetings.

Step one for a governing body is to simply understand what levels of technology are available (desktops, laptops, Chromebooks or tablets) and accessible for students and staff. Step two should be to assess their suitability and identify what plans are in place to mitigate any potential shortfall between the number of devices available and the total school cohort.

Drilling down even further, as governors we should be asking:

- How many devices do we have for both staff and students?
- Have we undertaken a survey to identify any shortfall if devices are needed off-site?
- Do we need insurance to cover this equipment?
- Have we been allocated devices from a national programme or will we need to purchase additional using our own financial reserves?
- Do all our students and staff have access to broadband or do we need to consider supplying 4G dongles (or equivalent) to support their access to resources?

Addressing the above questions should enable us to reach a point of confidence that all students and staff, when working remotely, have access to suitable technology with appropriate connectivity.

Device availability and connectivity serves no purpose, however, unless its use can be integrated into our teaching and learning plan for each child. The focus then is on adapting available technology to support provision for online education should it be needed, and so the following points should be considered:

- How do we evaluate the effectiveness of our current policies in terms of the provision of remote learning?
- Do we know what percentage of our cohort have access to our online systems?
- What alternative provisions have we made for those where connectivity is not an option?
- How do we ensure we are providing the right level of communication and feedback on learning activities for all children given their likely diverse needs?
- Do we have systems in place to facilitate effective communication between our staff as well as between staff and students?

- How are we shaping the mix of online education – the blend of synchronous and asynchronous learning?
- How can we continue to support staff and student wellbeing while working remotely?
- How are we supporting and developing parental engagement with our current technology use?

By this point, you will have already had plenty of opportunity to reflect on your school's delivery of remote education, but while there will have been lessons learned and successes achieved, we must also ensure those lessons and outcomes become embedded and form part of our future school provision. The rapid acceleration in digital skills required by all staff has been immense and the majority adapted remarkably well. However, we can quickly forget, and in order to retain confidence in digital technology, we must identify what elements we want to embed in the future. In other words, CPD must be at the heart of successful digital adoption, and the schools that adapted and were able to mitigate the impact of COVID-19 quickly were the ones that had already identified CPD in digital skills as an absolute priority.

Do school governors have a role in this part of the narrative? The answer is yes, absolutely – to objectively question and challenge, and ensure that we assess, reflect and drive the right strategy forward; they are the 'critical friend', providing the necessary support – be it financial or motivational – to facilitate or empower leaders to try new things to help shape the school's future.

Following on from the above, there are some additional questions that governors should be asking:

- Have we undertaken digital skills surveys with our staff, and are they being repeated so we can track confidence levels?
- Other than start-of-term inset days, how often are our staff receiving CPD training on the key tools we use within our classrooms and how are we developing this?
- Are we capturing suggestions and feedback from teachers on what does and doesn't work?

- How will we capture ideas from new teachers who bring experiences from their previous schools?
- Does our senior leadership team have practical digital experience or a good understanding about how technology can be effectively used in the classroom?
- Do we have plans for (or a review of) a digital strategy and how can we help?

I would also encourage governors to be outward looking. When faced with challenges, it is very easy to become inward looking – responding to your specific set of challenges and needs, basing your planning on the capacity and skillset available. There is an opportunity to avoid unnecessary pitfalls or short-circuit the process if we are prepared to look outside our own schools and seek peer advice.

For the last five years, I've been very fortunate to be chair of our regional governor leadership group – a group of school and academy chairs, briefed by the local authority and Ofsted on legislation changes and regional priorities, who also share challenges (and solutions) experienced in their respective schools. I have found it incredibly helpful to develop one-to-one and one-to-many discussions outside the formal structure of the leadership group, where we can peer-assess our own approaches to trust or school leadership and compare plans or strategies. Lately, that has of course included reflection on the measures we have put in place to support staff wellbeing and meet the challenges of a very changed educational landscape.

Distilling those collaborative experiences down, here are a few questions you should consider:

- Have we engaged with other schools or trusts to see how they have responded to recent challenges?
- How do our policies and approaches to governance compare to those of our neighbours?
- What worked best for other schools in our area in terms of online education delivery?

- Has anyone within our team engaged externally with other schools on their approaches to digital strategy?
- How could we benchmark and validate our effectiveness compared with our peers?
- If Ofsted (or equivalent) came to visit us tomorrow, would we have confidence articulating the decisions taken regarding delivery of online education and would we be able to evidence its success as a result?
- As before, I thought it would be helpful to conclude this chapter with some relevant articles that I have written in recent years which will hopefully help embed some of the topics we've just covered.

SECED MAGAZINE – AUGUST 2020

Making remote learning work

No school expected to be plunged so quickly into remote learning, with teachers and students separated from their peers and all learning from home. The learning curve has been steep – and it is ongoing.

On the plus side, thanks to increased amounts of EdTech and devices in schools, some teachers and students have been able to remain connected and learn within a virtual version of their in-school classes. We've seen how important this can be over an extended time in terms of maintaining a routine, trying to minimise learning gaps and even simply continuing the habit of learning when all other routines have fallen away.

What's different about remote learning?

Unlike a class environment where students can get support from each other and check things with their classmates, remote learning means that they suddenly need to become more self-reliant and proactive with their online questions to the teacher if they need clarification. This is a huge change and one that risks the students simply 'tuning out' if they don't understand. So lessons need to take this into account, as well as considerations such as screen exhaustion,[1] the fact that not every student has a computer and may instead be accessing lessons via a small phone screen, reduced concentration and a lack of supervision (especially needed to help younger pupils remain focused).

Teaching comes first

Although the technology needed to make remote learning happen is important, the teaching still comes first. According to a recent research paper by the Education Endowment Foundation, the thinking behind the structure of the lesson is the key: 'Ensuring the elements of effective teaching are present – for example clear explanations, scaffolding and feedback – is more important than how or when they are provided.'[2] Schools can help create more space for their teachers to focus on this more intricate mode of lesson planning and delivery by adopting straightforward, dedicated remote learning solutions that are easy to use, rather than complicated technology that could be a source of stress or distraction.

Be visible and reachable

A sense of connection is important for students' continued motivation and engagement, and technology provides teachers with a variety of ways to do this remotely. If their technology allows them to run 'live' lessons by broadcasting their webcam and audio out to students, that's a great start. But visibility can be more than just via webcam or video – the teacher can reinforce their connection to students by giving audio feedback, for example, or by using stickers or bitmoji and so on. It is also beneficial for students to have the chance to initiate contact with their teachers, so it may be that they can make themselves available in a 'breakout room' on a set day and time, where students can 'drop in' and interact.

Build independent learning skills

The increased tech-driven environment is actually a great opportunity to help students gain the independent skills they will need for remote learning, now and in the future. Teachers can model the use of calendars, schedulers and reminders so students know when a specific learning activity (e.g. a 'live' interactive session) is coming up or when assignments are due. They can also encourage the use of messaging and chat features to ask questions and share ideas – as well as teach the metacognitive strategies needed for resilience and perseverance, helping them to help themselves if they get stuck.

Peer-to-peer

Making provision for students to interact with their peers is vital. With nobody at their side to ask things like, 'What did Mr Smith say we had to do for question 3?', creating an opportunity for students to communicate not only provides that support, but means they can connect with their friends and classmates for a while, albeit virtually. Providing chances for peer interaction for activities such as discussions, collaboration, sharing and giving feedback can also go some way to sustaining engagement, especially when students are physically distanced.

Stay safe. Protect students. Practise good digital citizenship

Age-appropriate internet controls, filtering and context-based keyword monitoring are all devices used within schools to keep students safe online. And with the right technology in place on school devices, most of those protections should be able to continue as students use them at home. However, it is the teaching behind the concept of being online that enables students to develop good digital citizenship skills for themselves. Regularly reinforcing eSafety messages reminds them of the implications of their online interactions, as it is easy to feel protected from everything when separated from friends and peers while in isolation or lockdown at home.

The digital divide

A major contributor to exacerbating the achievement gap between advantaged and disadvantaged students is the lack of access to technology – and it is an issue that schools do not have the capacity to solve alone. Simply providing school devices to students will not bridge the gap if families cannot afford broadband, so it is here that remote learning needs to diversify to help ensure these students are not left behind. Whether it's spreading the word about where those with school devices can go to access free internet connectivity (e.g. libraries, community centres, etc.), pre-loading content onto devices to enable students to continue to learn offline, or providing paper-based learning resources for those with no technology access at all, planning and delivering lessons is a much more significant task than in 'normal' times.

Here to stay

Remote learning isn't a natural or ideal scenario for either students or teachers. But with the pandemic rumbling on and on, it's one many will need to adapt to and embrace. We are perhaps a bit late to the party, but blended and remote learning will likely be a part of the education landscape across the world from here on in.

References

1. RCPCH (2019). The health impacts of screen time – a guide for clinicians and parents. Accessed 14 August 2020 from www.bit.ly/3wkIlrt

2. Education Endowment Foundation (2020). Remote Learning, Rapid Evidence Assessment, London: Education Endowment Foundation. Accessed 14 August 2020 from www.bit.ly/3ox51Rz

PRIMARY SCHOOL MANAGEMENT (PSM) MAGAZINE
DECEMBER 2020

Building Primary Teachers' EdTech confidence

When we ask teachers how they feel about using EdTech in the classroom, at the same time as recognising that it is exciting, innovative, develops communication and creates learning opportunities, many also admit that they feel daunted and anxious about using it with their pupils.

It is easy to see why. If teachers are simply given a whistle-stop tour of a solution for an hour or so during an inset training day, they are unlikely to be immediately confident in its use. What they need is the chance to get their hands on it, practise and become familiar with it on their own terms. Without this time factored into their timetables, they will struggle to gain a working knowledge that enables them to use EdTech meaningfully with their students – and so the cost of purchasing and implementing it is wasted.

Where to start

Even if you do have a supportive SLT that is fully invested in helping staff develop their technology skills and has allocated time for you to do so, where do you start? Being left alone with unfamiliar technology can be intimidating, so receiving practical training that you can try out on the actual devices you will be using in class is crucial.

Accessing the technology as soon as possible after the training really helps to consolidate what has been demonstrated. Taking it slowly and becoming familiar with one feature at a time means that knowledge and confidence will build together, before you put things to the test in front of your pupils.

As with learning any new skill, repetition helps to achieve fluency. This rehearsal time is where making mistakes is beneficial as it provides you with the chance to find out how to fix things without being under pressure, thereby minimising the fear factor and leaving you better prepared for the classroom. Some teachers I have spoken with say they have practised by videoing themselves and, when happy with the results, have incorporated the feature into their video exemplars for pupils or parents. This is a really useful tip because, not only can you review and adapt as you go, but you will also build up a bonus library of instructional resources.

Four stages

A model for describing the stages teachers may identify with when learning new EdTech is defined by Mandinach and Cline (1992), who outline the stages of survival, mastery, impact and innovation.

With that mindset, if we hand a new solution to a teacher and provide little or no training, that places them in survival mode. They are not sure how to use it properly, are fearful of breaking it and, under pressure with 30 eager faces in front of them (either in class or sitting at home), confidence does not really come into it; it is just a case of whether they will sink or swim!

However, once teachers have learned the basics, they move to stage two, which is mastery. This is where they have received training and have had the opportunity to practise by themselves. They have also tried things out in lessons, and, when they have worked, this has begun to boost their confidence.

The reason that schools have invested in devices, software and (hopefully) CPD is that mastery evolves into stage three, which is generating impact. Teachers are no longer afraid of the technology, can cope when things do not go to plan, and they (and their pupils) are using it effectively.

The final stage that every school aspires to is to generate innovation. Here, technology is used intelligently and appropriately, teachers feel that they are digitally literate (with their technology knowledge on a par with their pedagogical and content knowledge (TPACK)) – so much so, in fact, that they are in a position to share those skills with others and, in effect, become the flagbearers for those less confident than themselves.

Practising and retaining skills

During the pandemic, through necessity, technology has taken centre stage. So whether collaborating and communicating in Teams, Zoom or Google Meet or helping students to learn via ClassDojo or Seesaw, many teachers have worked hard to significantly raise their EdTech skills in a short time – and for that, we applaud you!

What is critical, though, is that these new-found skills are not lost once we begin to move past COVID, when the urgent need for remote teaching and learning inevitably diminishes. For that not to happen, the progressive use of EdTech needs to be embedded across the school. Schools can ensure this by reviewing and standardising their solutions, thereby making things easier for staff moving between sites within a Trust, and easier to support. So deciding, for example, whether you are an Apple/Google/Microsoft school is key and gives you the foundation on which to implement complementary applications that are therefore more accessible (in terms of intuitive usage) for your teachers.

Continued learning support is a fundamental part of retaining any new skill. This can take various forms, such as ongoing formal CPD training sessions, top-up/revision training, peer sharing, solutions champions or interacting on dedicated online forums to ask questions and share answers and experiences with others. The key is to keep your knowledge ticking over and evolving with changes in the technology, rather than letting your skill level drop and having to play catch-up. This way, you will retain the knowledge and confidence to use EdTech as a tool to innovate, rather than simply just 'use' it.

Future investment

There has never been a more important time to be digitally literate, and the pandemic has been a huge catalyst for change in this respect, with the need to teach children remotely and maintain communication with parents to support the continuation of learning. As many schools will maintain their current EdTech use after COVID, the work teachers are doing to increase their digital confidence now will integrate technology into their teaching practice, so that it moves from being a box they must tick to being a tool they automatically use to achieve their pedagogical aims.

Getting to grips with EdTech

Learn at your own pace. Rushing things means you won't take information on board fully and increases the likelihood of coming unstuck later on.

Don't be intimidated if others appear to learn faster. Stick to what works for you. You know you'll get there eventually.

Testing, testing. Try things out with an audience of one: yourself. Video yourself and watch it back. You'll soon see which bits need refining.

Mental preparation. If you have the sequence of what you need to do outlined clearly in your head first, then the practical side will follow.

Take it steady. Minimise stress and validate your progress by introducing just one new tech feature into your lessons at a time.

Giving is receiving. Share experiences and tips with colleagues in the same situation as you and they'll reciprocate. You'll all pick up some great ideas.

References

Mandinach, E. B. and Cline, H. F. (1992) The Impact of Technological Curriculum Innovation on Teaching and Learning Activities.

EDUCATION EXECUTIVE – NOVEMBER 2020

What impact has this year had on #EdTech use in schools?

Now that we are a few months further into the pandemic and into our second lockdown, it seems a good time to evaluate the changes that have occurred within education. As we settle into the new reality of using the online space to enable the continuation of education amidst disruption to school time, it is useful to look back and see just how far we have come.

First lockdown

As we know, at the point of the first lockdown, not all schools were on a level playing field when it came to technology use. Some had already incorporated lots of EdTech into their teaching and learning, which made for an easier transition to online-led education. Others, however, were on the back foot and had to try to catch up very quickly.

With the lockdown came an evolution of the term 'blended learning'. Up until that point, blended learning had meant students being taught in person in the classroom, with opportunities for EdTech interaction being incorporated into the lesson, providing them with a mix of synchronous and asynchronous learning activities. With teachers and students now remote from each other, the term took on a new definition and the 'blend' became a mix of in-school and remote (or 'at home') education. The synchronous element is now the real-time technology-based teacher interaction with the students, and the asynchronous component is provided in the form of pre-recorded video exemplars, resources for project work for self-study and so on.

Another hurdle to overcome was that of teachers' EdTech confidence. Many were new to the concept of teaching with such a heavy emphasis on technology and had to work hard to not only learn new solutions at breakneck speed to maintain contact with students learning at home but also adapt their lesson plans for the online arena. Without prior experience on this scale, some got the balance right and some did not, which was only to be expected.

Of course, not every student has access to technology at home and not every home has the connectivity or devices needed to simultaneously support one or more students learning online, plus parents working remotely. The virus certainly highlighted the digital divide – and that was something that was not (and is not) easily solved.

What is different this time?

The main change for the education sector during this lockdown is that there is a clear government mandate that schools and universities should remain open. We should remember, though, that schools never completely closed during the first lockdown; they were still open for vulnerable children and children of key workers. So actually it is not the case that schools are 'staying open'; it's more that the aspiration is that all students will be in school. However, the likely path over the next few months will be increased cases of COVID-19 within student bubbles and so the blended learning model of in-school and remote teaching will come into play.

This time, though, teachers' confidence with technology is much improved. Once schools selected a platform to work with, teachers dedicated themselves to learning how to use it and getting to know the possibilities it offered for their online lessons. Nevertheless, there is still more to be done, and for many schools, supporting their teachers to learn and be confident with EdTech will rightly become an ongoing process. It certainly has highlighted the need for CPD in this area.

Despite 'staying open', for many schools, a reduced timetable is a real possibility this winter (staff numbers permitting), with students attending for just a couple of days a week, thereby allowing schools to significantly reduce the density of children in the building and minimise the virus transmission risk.

From our own and others' experiences during the first lockdown, we are all now much more conscious of the importance of looking after wellbeing and mental health. Providing face-to-face connectivity for students and teachers can mean so much – and it really helps those students feeling isolated at home to maintain their connection to their teacher and to learning. It's also really important for schools to facilitate peer-to-peer collaboration because we know there is much value in children being able to connect and communicate between themselves as part of their learning journey.

What will EdTech use look like in the future?

The coronavirus has been the greatest accelerator of change in school EdTech in recent times. Schools have already learned how to use appropriate solutions for teaching and learning, embedded their use in lessons and increased their user confidence, enabling learning to continue in new and interesting ways.

Some of the technology and how it has been used during the pandemic has already created positive change. A good example is the move to online parents' evenings. Schools have found it is much easier to manage fixed time slots online for each set of parents/carers to avoid cutting into the next parents' time. In addition, parents actually prefer having a private conversation with their child's teacher, rather than being in a room full of others. The move online is also much easier for working parents, who can drop in for the required time with minimal disruption.

The online tools and techniques schools have used during this time will be just as appropriate for exam revision sessions during the Easter holidays, saving teachers and students from travelling into school. Schools will need to ensure though that students can either connect live or access recordings of those sessions subsequently, so nobody misses out.

Some of the schools I have talked to have also noted that the use of technology has significantly improved their engagement with parents – and particularly with those who have traditionally been harder to reach – which is enormously positive. After having to support their children's learning at home during the first lockdown, many parents have a new-found appreciation for the job that teachers do, and I sincerely hope the post-COVID world sees a continuation of that!

From a Vendor's Perspective

There is an abundance of resources and articles available that cover every aspect of using EdTech within schools, and naturally they will have a finite lifespan and a need to constantly change and adapt in response to market demand and customer expectations. Being heavily involved in the EdTech industry, what is obvious to me is the lack of information specifically from the perspective of an EdTech vendor.

It would be easy to assume that the main driver for any EdTech organisation involved in developing products or services is profit. Whilst profit is a key driver for most EdTech vendors, to assume that this is the only consideration would be to miss a host of other factors that help to shape decisions on an approach to market. I am sure that there are many educators who have created EdTech products and transitioned into the commercial world, and whose motivation wasn't solely profit-related, so clearly there is a sliding scale here too.

Let's not be naïve – generally speaking, the larger the entity and the less involvement on a personal level, the more the numbers will shape the focus. However, to create profit, any successful vendor must understand a much broader range of variables to mitigate the investment risks taken when producing something new. In addition, the more agile the target marketplace, the narrower the window of opportunity to make that return. Much like the duck shooting challenge at a fairground where you get one shot to hit the mark as they move past, if a vendor's proposed solution is designed to meet a specific need, there is usually just one shot to get to market at the right time – either before another vendor swoops in and takes the customer base with their new solution first, or the market changes and the product need no longer exists. Although not EdTech specifically, the most extreme example of this were the tools developed to mitigate the **Y2K bug** on devices. Clearly, launching on January 2nd 2000 was never going to result in a successful outcome.

To add further context within the broader narrative of how EdTech products can have a positive impact in schools and how they may shape our plans, I feel it would be helpful to share some of the thought processes and strategic considerations that an educational technology vendor needs to make. Specifically, how that shapes the products produced, how to best engage with the marketplace, how the COVID-19 period may have changed a vendor's focus and view of the landscape, how all of that may help educators approach and engage with vendors, and what they need to consider when looking at solutions in the context of the vendor and their broader ecosystem.

As a vendor in the EdTech space, I guess it's only fair to share our journey at NetSupport in terms of the approaches and decisions we've taken and how they translate into the products you see and use right now.

What vendors want

Just like students, no two vendors are alike. Each will have their own aspirations, visions and values, but that distinction becomes even wider within the education space as some vendors are personally invested, so the desire to make a difference isn't always measured just by the bottom line of profit and loss. For some, it's about creating capital to fund further enhancements and develop new solutions.

In the same way that vendors need to understand education and the needs of their customers, I believe it's equally important to explain and understand the vendors' perspective so that we have a more informed set of benchmarks and criteria when reviewing suitable solutions, particularly when that sometimes involves partnership or co-production with a vendor.

This section aims to provide more context to the topics covered so far in this book.

The strategy

There are two parallel (and fundamental) processes that shape the formation of an EdTech company's strategy when creating a new product, each of which covers a wider set of variables.

The first process is proactive and involves analytical and statistical considerations: understanding the market growth potential; segment size; customer expectations; competitors and technology trajectories. The information gathered through this process helps to shape and define the start and end of a development project.

The second process is more reactive and tends to occur once the project is underway, reflecting changes in market shifts, operational decisions, ongoing financial costs, capacity, internal ideas and so on. The decisions within this process tend to be less strategic but vital in further shaping the original strategy.

These processes are defined by '**The Innovator's Solution, creating and sustaining successful growth**' (Christensen and Raynor, 2003, p. 215) as the deliberate and emergent strategies – the constant reshaping of original plans (market and competitor analysed) based on the evolving input from operational factors.

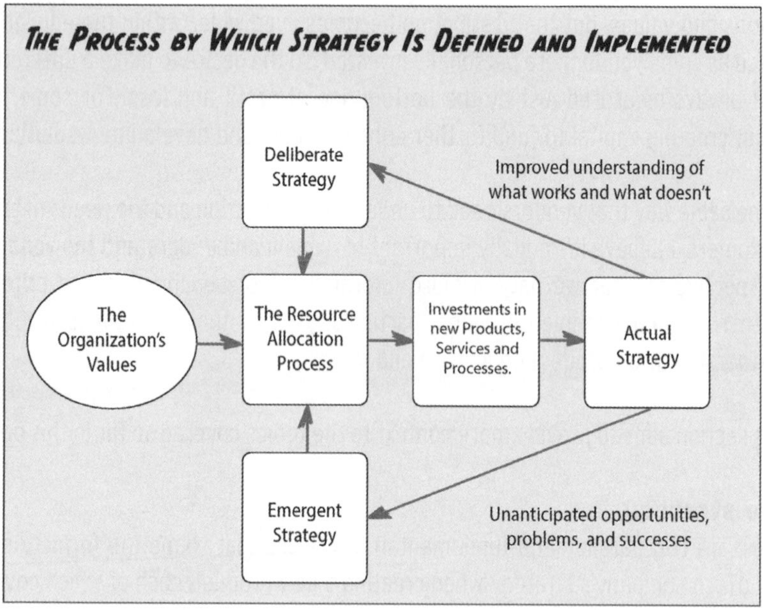

This is precisely the journey we experience at NetSupport – we start a development plan with a clearly thought-through 'road map', often beginning with a very heavily narrated whiteboard; then during the development process, new ideas are formed, different ways to achieve an outcome are analysed, competitor announcements and customer feedback are considered – all these elements influence the finished solution. It is far from a fixed process but what is key to understand, as an educator, is the vital role the customer can have in shaping and adding value to a solution – not just after the product is released but whilst it is being developed. Good vendors are smart and figure this out quickly.

So, what are the considerations if you are a budding edupreneur?

I've been very privileged to have met many educators over the years, and every so often they will express an idea or a solution that they think would benefit their classroom and seek advice on how they can turn that idea into a reality. Although

I willingly share my thoughts and suggestions wherever possible, it will probably come as no surprise to know that whilst it is easy to develop a solution, it's not that easy to develop a *successful* solution – the failure rate for new products and services can be as high as 90% in some sectors.

I never profess to have all the answers, but I'm happy to share my 30 years' experience along a well-trodden path of lessons learned from making a cornucopia of mistakes (or 'opportunities for development') alongside a few successes. To be successful in business, you have to be willing to take risks, know that not everything will pay off, but believe that even the failures will teach you lessons that will help fine tune your knowledge and approach next time round.

We can probably define the 'edupreneur' process as identifying an opportunity, getting a sense of its potential, developing the solution, validating (testing) and then launching. It is, of course, a process I and the team at NetSupport have been through many times and I read a great article in Forbes (Toro, 2016) by **Juan Manuel de Toro**, from the IESE Business School in Madrid, who condensed the whole process down into five steps (www.bit.ly/2T5MaS1).

1. **Identify an opportunity and generate a new idea to fill it.** That sounds painfully obvious as clearly if nobody wants or needs your product then the odds are it's not going to be a success. There are rare exceptions to the rule, when some of the largest technology vendors take a strategic position to create an entirely new need (and product) and then attempt to convince their customers of its benefits and why they must have it. This approach requires scale and a significant marketing budget so for the purposes of this section, let's just stick to the basics that for the most part, there must be a clear 'need' for any solution you plan. You have to think about current as well as future customers – how happy are they with alternative solutions already provided by competitors and what is the potential to improve and innovate beyond the existing solutions?

2. **Measure the opportunity.** Once you are satisfied that there is a genuine opportunity, you need to need to reflect on current market trends (perhaps

an area where it is anticipated or expected that there will be greater investment), consider the shifting landscape in the platforms that customers are using (for example, delivering a curriculum that only runs on Linux might somewhat limit your potential). You can also use a more formalised method, such as the **Kano Model** (www.kanomodel.com), which categorises five different ways of understanding and prioritising customer requirements for a new product or service.

3. **Develop the concept.** This is the equally exciting and stressful part – developing the product, the key focus being on refining your idea, making sure it will meet the needs of your customers and, in a competitive space, stand out from the crowd. This is where vendors who embrace the co-production approach can have an advantage (I'll talk about this in more detail further on). It's also reassuring during this phase to get validation and feedback from the coalface and gather ideas for extra features you can add to enhance the product even more.

   For over 18 months, we did exactly that while we were developing our **ReallySchool** solution (www.ReallySchool.com) – primary assessment and learning journal creation – with the EYFS (early years foundation stage) team at Dogsthorpe Infant School. During the later stages of development, we included other schools to help validate the product even further. The support of and input by Rachel Jordan (EYFS lead) and Becky Waters (headteacher) was invaluable and represented a win-win for everyone – it made a huge difference to the final product and they got exactly the solution they wanted and needed.

   By this stage, the solution is nearing the end of the development phase and is hopefully going as planned, with constant checkpoints and feedback that hopefully allow us to meet the three key objectives: having a solution that will meet (or exceed) the needs of our customers; having functionality that will make it stand out from the competition; and having the potential for a profit. The next stage is vital.

4. **Testing, testing.** For all the great EdTech products out there, there are some that are...flaky, or have clearly been developed without the foundation of a rigorous testing process. No software vendor could (or should) claim their product is 100% perfect because that is virtually impossible given the constantly changing market they are operating within. Updates to operating systems alone very often 'break' functionality that software relies on. But regardless, the testing phase will have a significant impact on the success of any solution. This phase should include key considerations for the UI (user interface) and how accessible the solution is to use, performance of the solution for any potential overhead on system resources, and (if cloud-based) suitable security and penetration testing to ensure customers' data will be protected. In the latter stages of the development cycle, there is an opportunity to consider beta testing (sharing a pre-release version with selected customers) for validation that the product operates as expected in a real-world environment. This process becomes a form of co-production if the right systems are in place to capture and act upon any feedback received.

5. **Position and launch.** This stage is about making sure the final product is positioned with a clear message that ensures customers have the correct perception of its benefits, and at a suitable price point. The selling price will vary depending on organisational aspirations, market share, speed of acquisition versus initial viable profitability. In other words, the lower the price, the quicker the product will sell, the more market share will be gained and further investment by competitors discouraged – but make sure it's sustainable in the long term.

   From an educational perspective, of course, the message is always that funding is tight so ideally, products should be priced as low as possible. In theory that's fine, but there must be a balance – to achieve a product that is well developed and continues to evolve and meet customer need, it must be adequately funded. On a personal level, we rarely choose the cheapest option when deciding on purchases such as clothing, cars, furniture, etc. We consider many different factors before making an informed choice, and the same logic should be applied to choosing the right software solutions. There is, of course, a balance – it doesn't matter how amazing a solution is, if a school can't afford it, then both parties miss out on an opportunity. Pricing for education will (or should) always be keener than for other sectors.

In summary, the considerations on your vendor checklist should be:

- Identify the opportunity.
- Evidence or estimate the demand.
- Consider the environment (platforms and possible changes in landscape).
- Review the competition.
- Identify the potential longevity of your solution.
- Consider the overall development costs to reach the market.
- Research and forecast the ongoing operational costs.
- Be clear on the product's purpose and your intent/expectations.
- Undertake testing throughout the process and validate that it still meets customer need.
- Position and launch the product with a clear narrative and plan.

As a final note, also consider the potential opportunity cost – what else could you be developing instead of the current solution and what opportunities might be missed by not doing so?

On the subject of costs, at this point I should also mention a particular operational cost that needs to be considered as it can often catch a vendor out – namely, your cloud-based hosting platform, e.g. Microsoft Azure.

Until 2010, software solutions were relatively straightforward when deployed locally on a school's LAN or WAN or, occasionally, a vendor's own central servers. In 2010, the landscape changed significantly with the introduction of Azure, a cloud computing service that provided a platform for vendors to build and deploy solutions through Microsoft's managed data centres.

Satya Nadella, CEO of Microsoft, recognised that Microsoft needed to further develop their offering for cloud computing and saw that this was an ideal opportunity to significantly grow their revenues by elevating cloud services as a core priority for the business. He was certainly proven to be correct in his judgement, given that by 2018, the Azure platform accounted for $20 billion per annum in revenue for Microsoft. With two other dominant names in the space – Amazon Web Services (AWS) and Google, Gartner predicts worldwide public cloud spend will grow by 18% in 2021 (https://gtnr.it/3bIGL9L).

By February 2021, **Canalys** data reported that AWS had 31% of the market, followed by Azure at 20% and Google at 7%, with Alibaba cloud close behind (www.bit.ly/3hJztGz).

The reason I am including this here is that a good slice of that revenue comes from vendors developing cloud-based solutions in response to the push for cloud computing, and it has now become a more expensive proposition for vendors to offer solutions to their customers, so inevitably customers are also having to pay more for those solutions. To clarify this even further, vendors not only pay a fixed amount for 'virtualised servers' (or services in the cloud) but are also charged transactionally for every API call, data upload/download, and the amount of data kept in storage. The more a cloud-based system is used, potentially the more it costs the vendor (unless they host their own cloud services). Understanding the implications of this is key when developing a solution so there is full awareness of the cost impact with each layer of functionality added.

Vendors should also consider their data retention policy and be mindful of streamlining all processes where possible. When considering a cloud-based solution, educators must have confidence that the solution is viable and that they are fully aware of any hidden costs that may be incurred through additional usage. Finally, although I will mention it again further on, schools must ensure they have undertaken a data protection impact assessment and know exactly what data is being shared and hosted online, where it is and who has access to it.

What if you're a vendor with existing solutions?

As I reflected earlier, most technology is an evolution of earlier technology or approaches, rather than a revolution. In the same way, for many established vendors in the EdTech space, the next step forward is usually the evolution of an existing product. Often referred to as sustaining technologies, the purpose is to improve the performance of established products, but when disruptive technologies emerge that cause a more significant shift, in terms of functionality, they can take a step backwards in favour of offering other benefits that might be more appealing to their customer base: improved accessibility, online delivery, integration options, etc.

This creates what Clayton Christensen (2013) refers to as '**the innovator's dilemma**' – do you keep moving your current solution upwards, adding more functionality, or do you re-enter the space with something new and disruptive?

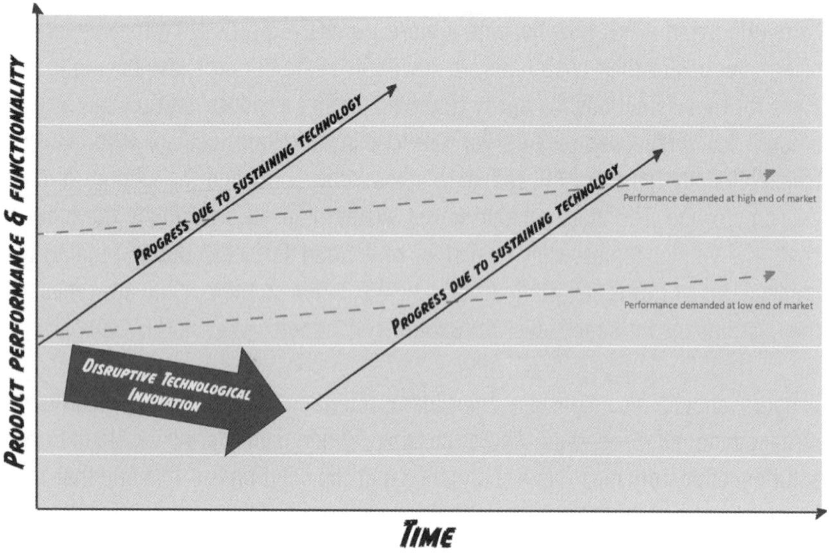

In simple terms, as your product evolves and you continue to add features, often in response to competitors, your solution changes from meeting the basic features expected by your customers to exceeding the limits of what they need (the first line in the chart above). Continually adding features to your solution, like diminishing returns, isn't necessarily always to the overall benefit of the solution or the customer. (Anyone who has had to re-learn several successive versions of Microsoft Word will be familiar with this phenomenon!)

As an alternative, you could disrupt and create a new solution with a new format and platform, just as we did at NetSupport by offering a new cloud-based classroom solution that could flex for remote learning (www.classroom.cloud) alongside a mature desktop equivalent. In doing so, the line is then reset – providing basic functionality initially (but in a new accessible format) and then over time, based on customer need, adding more features.

There is another crucial lesson here – sometimes less is more. Accessibility, usability and price point are often more important drivers for schools when adopting new technology than who has the longest feature list.

Co-production

Education is a very agile space with multiple layers, and the assumption that solutions can be developed and automatically fit customer need without some form of collaborative process is not just a missed opportunity: it's counter-intuitive. If I were a maths teacher with 15 years' experience in the classroom and I decided to create a maths app, it's reasonable to assume I might know exactly what my immediate peers and students needed for my app to have impact. But if I wanted to scale up and become a mid- to large-size vendor, the decision makers aren't always those at the coalface with the best ideas and suggestions. No matter how good your idea, you need real-world validation, and recent events have shown us how quickly priorities and challenges within the classroom can change. The benefits of agile development are therefore pointless without equally agile feedback loops running alongside.

In May 2020, I wrote about the subject of co-production for **Headteacher Update magazine** (www.headteacher-update.com) because I wanted to encourage vendors and schools to embrace and support the concept that the future of technology in education needs to be driven by collaboration. After all, this was one of the key takeaways from the Department for Education's digital strategy, launched in 2019, and I am sure anyone involved in education would agree.

Educators are particularly good collaborators – with each other, with government initiatives or with parents and students. However, when it comes to collaborating with commercial suppliers, school leaders can be reluctant.

Co-production – vendors and schools working together to develop solutions – can be a challenge. Done properly, it can benefit the school as much as the vendor, gaining preferential and cost-effective access to state-of-the-art technologies. Done badly, it can be resource-draining. In my position as a vendor and a multi-academy trust board member, I see co-production from both sides. But irrespective of which side you sit, you should consider it.

Any technology used in schools must be underpinned by evidence that it does the job it's supposed to do. Without evidence of an EdTech solution's impact (derived from in-school testing), we stumble into challenging territory – buying an unproven solution or disregarding best practice in choosing solutions that have been thoroughly considered, researched, and tested.

Groups like the **EdTech Evidence Group** (www.edtechevidence.com) advocate that schools need to be able to easily assess the value and impact of products and services, and whilst I support them in their broader recommendations, I'm conscious that it's far from easy to find common measures of evidence – a 'one size fits all' checklist approach to validation – simply because of the breadth and diversity of solutions that fall under the educational technology umbrella. (In the next chapter, I share some ideas about 'asking the right questions' to ensure purchases are appropriately research informed.)

If schools are to benefit from evidence-based technology, they must give companies the opportunity to engage with them during the technology's development phase. Otherwise, schools would be limited to only choosing existing, fully established solutions with proven evidence behind them, and this creates a real barrier to the innovation and development of new solutions.

As the DfE digital strategy states:

> EdTech businesses often struggle to access education institutions to test, pilot and prototype their products. Teachers, school and college leaders and business managers are rightly focused on the day-to-day needs of their students and their organisations. This means the feedback that EdTech products and services developers receive can be limited, which in turn hinders the ability of EdTech innovators to evaluate and refine their offer.

Despite the government's focus on greater collaboration between educators and EdTech companies, there is a lack of advice on what this should look like. As part of the DfE's digital strategy, some schools are acting as testbeds for co-production best practice, but it's very much early days.

What are vendors' real intentions?

I've already explained the value to a vendor in reaching out and fostering relationships with schools during their product's development phase, but let's now briefly consider the same topic from a school's perspective. Co-production can be rewarding if both parties are clear on the terms of their relationship and for schools who may be keen to work alongside vendors, there are several issues to bear in mind.

If we all agree that it's fundamental to ensure vendors deliver what teachers need, rather than what they *think* teachers need, when it comes to co-production, knowing whether the vendor is truly entrenched in the education community really matters. This means looking beyond the marketing brochures and websites and questioning the people behind the products. Who are they and where do their interests really lie? Do they understand and have proven experience within education? Are they putting time and effort into understanding educational needs or are they on the periphery, simply selling into schools? It means asking probing questions: finding out about their background, any external roles, what evidence and research they can provide to support their approach. Of course, I'm a little biased, but to me an EdTech vendor without anyone from education is, well... just selling the tech. It's easy to be drawn in by the latest trending technology buzzwords, but you need to be confident that their business is as invested in education as it is in technology, or at least on a journey towards that.

It's also important to be clear about the boundaries of the project. Is this co-design (a long-term exploration of the needs and opportunities for tech within the school) or is this just a 'test and learn' situation (where vendors work with schools to rapidly test their product prototypes so they can improve them)? The latter is the more common arrangement.

As an educator, if you are considering a longer-term co-production relationship with a vendor, then it's essential they do not isolate technology from teaching – they must consider the whole context. That means they need to be as focused on working with teachers on pedagogical development as they are on technological innovation. Again, it's important to know the people behind the product and whether their interest extends beyond technical processes to the broader pedagogy considerations as well.

Does the project align with your school's EdTech plans? Before engaging with any EdTech product provider, schools should look at their EdTech plan and reshape it under the broader umbrella of your digital strategy (which, at its simplest, is a plan of what you want to achieve and why) and then assess whether co-production will pay off. There is little point agreeing to put effort into testing an application which does not result in any impact where it's really needed.

You may be offered any number of incentives to test a product but remember, just because something is free doesn't mean it's cheap. The main cost of co-production is usually time – implementation, training, feedback – and it all adds up. If that cost cannot be attributed to a long-term key objective, it's pointless.

Any decision regarding technology should be made with the school's digital strategy in mind. It might be called something different, or it might be a range of interwoven plans embedded within different parts of your school's overall strategy rather than a single plan, but as the DfE's advice for schools seeking to devise a digital strategy is a little lacking, I have covered this in more detail in another chapter. (It's almost as if I planned it this way...)

EdTech vendors need to be working with schools to find out what schools really need from a product. There have been so many times when I've been given the hard sell on a product that claims to be the answer to all my problems, but when I ask about the reporting side of it and if it can produce detailed reports on user statistics, I usually get told that it's in development and that will be the next thing they look at. Why not work with schools and leaders to begin with to really find out what we need, rather than assuming (from a very limited knowledge base) that what they have created is what we need.

Jon Tait, Director of School Improvement and Deputy CEO, Areté Learning Trust

Digital disruption

Digital disruption is a transformation caused by emerging digital technologies and business models. These innovative new technologies and models can impact the value (or perceived value) of existing products and services offered in the sector. Think Netflix – and just what that meant for Blockbuster Video!

Another example – what Amazon did to the well-established book-selling market in 1994. I read somewhere that Jeff Bezos originally intended to call his online business 'Relentless' until friends told him it sounded sinister, so he chose Amazon as it was a place that was 'exotic and different'. (As an aside, even if you head to www.relentless.com now it will actually take you to Amazon.com.)

The term 'digital disruption' is used when the emergence of new digital products or services disrupts the current market and causes re-evaluation. In an earlier section, I shared the innovator's dilemma – you either keep adding to your current solution (more shelves in the bookshop and a wider choice of topics) or you disrupt and offer something completely different (e.g opening an online book shop instead) but at the sacrifice of choice to start with (no need to leave home to get a new book, but initially with a smaller choice of topics online, while the new business scales and establishes distribution).

We might argue that interactive whiteboards were a disruptive technology (and the jury is certainly split on the long-term benefits they have delivered), but what I wanted to flag was how a disrupter retains its position in the market once established. In 2020, I was fortunate enough to undertake a three-month course on '**Digital Disruption in the Covid Era**' at the Judge Business School within Cambridge University (www.jbs.cam.ac.uk), and the most relevant strand that I wanted to explore was about the strategies behind building ecosystems. I will expand in a moment, but in essence, it's about how to make your offering 'sticky', so a customer is less likely to go elsewhere.

From a vendor's perspective, it means that instead of having one good quality solution, we need to develop a range of integrated solutions, the sum of which not only provide even more value to their customers but also provide the convenience and benefit of different, but cohesive, solutions that hopefully mean customers are happy to stay with their vendor.

It sounds straightforward, but it has pros and cons from a customer's perspective.

If I create an Amazon account and purchase all my eBooks and music through them, it makes life simpler and more convenient with a single sign-on and purchase account. Amazon's solution is compelling, but it is also sticky: it becomes a painful process if I choose to leave and go somewhere else if there is a better supplier of books or music online.

That retention is a real positive for vendors, of course, who want that year-on-year repeat business.

Flip the concept 180 degrees and the same considerations need to be made by schools when choosing their solutions. If you sign up to this school information management system or that online portal for student assessments, will it make it too difficult to switch to another solution in a few years' time? Is your current solution so sticky that it will prove to be a nightmare trying to move your records and data to another solution, or will it disrupt functionality that you still want to use? Does that place you at risk of overpaying and being locked in if the vendor changes their pricing model further down the line?

In our enthusiasm to embrace a new all-singing, all-dancing solution, schools need to be mindful of whether adopting it might compromise their choices in the future – and with the fast pace of development and innovation, it's a given that there will always be other choices. At the same time, vendors also need to focus on that stickiness and how it can be used positively when developing **quality** solutions. The two aspirations don't need to be mutually exclusive, especially if you co-produce.

Post-COVID-19

As I have said throughout the book, the focus and opportunities that lie ahead should be about widening the lens we look through, in terms of reflecting on what has and hasn't worked. Even the failures are useful stepping stones to success. Looking ahead to post-COVID-19, as a vendor, in addition to seeking insights from the long-standing and meaningful connections we have with different educators throughout our markets, I have found it helpful to keep the view as wide as possible to gauge where we could, or should, be heading.

I read a great article entitled '**What will education look like in 20 years?**' (Schleicher, 2021) by **Andreas Schleicher**, from the Directorate for Education and Skills at the OECD (A. Schleicher, weforum.org, Jan 2021) and he rightly identified that imagining an alternative future for education is a key catalyst in pushing us to think through different outcomes and help new agile and responsive systems develop. The focus of the underlying question was the extent to which our current spaces, people, technology and, of course, available time are helping or hindering the creation of a long-term vision. To give a very condensed summary of the topic, should we be looking to add some polish and enhancements to our current systems or come up with an entirely different approach to the way we organise our resources and deliver education? He posed four possible scenarios in the future of schooling:

1. **Extending schooling** – participation in education continues to expand; international collaboration and technological advances support more individualised learning; the structures and processes of schooling remain.

2. **Education outsourced** – traditional schooling systems break down as society becomes more directly involved in educating its citizens; learning takes place through more diverse, privatised and flexible arrangements, with digital technology a key driver.

3. **Schools act as learning hubs** – schools remain, but diversity and experimentation have become the norm; opening the 'school walls' connects schools to their communities, favouring ever-changing forms of learning, civic engagement and social innovation.

4. **Learn as you go** – education takes place everywhere, anytime; distinctions between formal and informal learning are no longer valid; society turns itself entirely to the power of the machine.

(OECD, 2020)

Right now, option (1) seems the most likely and, frankly, achievable outcome. But I can see elements from all four in the future, particularly as option (4) is already evolving within the further education space. But whether we want to judge this as big picture thinking, educational disruption, or just pie in the sky, the reality is that we are on an inevitable pathway to change, and either we hang on and discover where we end up, or we proactively plan and help shape the direction of travel.

Vendors have a key role in this process – not to dictate or decide, but to support, facilitate and, most importantly, provide options that allow concepts which are (currently) considered unachievable to become a reality. There are elements within all four options above that could be aligned with some of the creative activities our schools, trusts and districts have undertaken during the COVID-19 pandemic, and whilst not all have been successful, given the speed of implementation, they at least showcase that our original physical definition of the school has begun to change irreversibly.

As technology vendors, we need to have our fingers on the pulse – not just in developing tools that respond to current needs within the classroom, but also in undertaking research and being engaged in the broader blue sky thinking to validate and provide insights into what may or may not be achievable as we move forward. If I place my commercial hat firmly on my head, where there is change, there is opportunity; and where there is opportunity, any good business should want to be involved.

I'm flying the UK flag here for all new start-ups and established technology vendors – if you aren't already a member, join **BESA** (www.besa.org.uk). They represent the entire UK education suppliers' sector, help focus the collective voice, provide industry research and insights and are a great resource to grow within your market or when looking to develop or extend your export activities.

> Through their work with schools and education systems around the world, EdTech vendors have a wealth of knowledge of and advice for what works well in schools. EdTech suppliers are keen to work hand in hand with schools to share knowledge and support schools. EdTech vendors must continue to have access and opportunity to work with schools collaboratively to help develop solutions to new and emerging challenges facing schools in the future.

Caroline Wright, Director General, BESA
(British Educational Suppliers Association)

PRIMARY SCHOOL MANAGEMENT MAGAZINE – SEPT 2020

Is there a place for technology co-production in primary schools?

With the UK's education technology sector set to be worth a whooping £3.4bn in 2021, there are now hundreds of technologies being targeted at primary schools hungry to take advantage of innovative teaching aids and admin-cutting tools. But are all these technologies made equal? Of course not.

My advice to primary leaders considering any new technology is to ask what role schools themselves have played in its development. Co-production – suppliers and schools working together to develop technologies – demonstrates that, at a basic level, the technology answers a real need and takes into account direct teacher feedback.

For ambitious primary schools, keen to innovate, engaging in co-production projects can return preferential and cost-effective access to state-of-the-art technologies. However, such projects done wrong can be a resource-draining exercise. In my position as a vendor, multi-academy trust board member and primary school governor, I understand both sides.

Co-production can be a rewarding exercise for both school and supplier if both parties are clear on the terms. For schools considering such a partnership there are questions they need to consider several factors.

What are the vendor's real intentions?

In education, co-production is fundamental to make sure suppliers deliver what teachers actually need, rather than what they think teachers need. Therefore, whether the vendor in question is truly entrenched in the education community really matters.

This means looking past the marketing and questioning who the people behind the product are and where their interests really lie. Do they have experience within education – putting effort into understanding educational needs – or are they on the periphery, selling in?

That means asking probing questions: their background; external roles; what impact, evidence and research do they have to support their approach? You need to be confident that they are as grounded in education as they are in technology.

If you're considering a longer-term co-production relationship with a supplier, it's essential they don't isolate technology from teaching. That means they are as focused on working with teachers on pedagogical development as they are on technological innovation.

Again, this is why it is so important to know the people behind the product and know whether their interest extends to pedagogy beyond technical processes.

Does the project align with our school's digital strategy?

Before engaging in any EdTech project, schools need to ask, 'Does this align with our digital strategy?' If your school hasn't defined its digital strategy, then be very wary of committing to any EdTech project!

In my experience, primary schools tend to lag slightly behind their secondary colleagues in setting a digital strategy, mainly due to the perception that it requires a large IT team to do so. In fact, a digital strategy is a joint effort that, at its simplest, is a plan of what you want to achieve and why.

With this in hand, you can then assess whether co-production will pay off. There's little point agreeing to test an application which doesn't result in impact where you need it. You may be offered any number of incentives to test a product but remember, just because something's free doesn't mean it's not expensive.

The main cost of co-production is usually time; implementation, training, feedback all adds up. If that cost cannot be attributed to an identified, long-term key objective, it's a waste.

While devising a digital strategy can seem like a daunting task there are guides and expertise available to help.

What should we expect and how closely to engage?

Every co-production project will be different. It could involve working with an existing supplier on current solutions, testing new or refined features. Teachers will already be using the product, know its strengths and weaknesses and may already have clear ideas for improvement and evolution.

Alternatively, it could be beta testing a product, putting a solution through its paces prior to release, providing early access to discounted or free versions of new technology, which can be exciting.

It might be starting at the drawing board and helping shape a new solution to meet a current need. However, testing or collaborating on specification can create extra work for teachers who need to be trained and a commitment to reporting and providing feedback.

That's why it's so important to reflect on your digital strategy.

Will the time and effort required get us closer to our goal? If it aligns then make sure you understand and agree with the level of support and training teachers will be given, the mechanisms to enable fast, streamlined feedback and the added extras you'll access as part of the scheme.

Don't be afraid to ask the question to yourselves – 'Do we get as much out of this relationship as we put in (or more)?'

Don't be guinea pigs; be partners

Let's be clear. Co-production is about shaping and polishing solutions so they are the best they can be. That means suppliers must be transparent about why the technology exists and any issues they are seeking to solve.

With that in mind, co-production is not about putting ill-designed, half-finished products in the classroom, using students as human guinea pigs! No supplier should expect a school to risk student attainment or teacher wellbeing in this way – if they are, walk away.

Seek a co-production partner that is open and honest, understands and appreciates your digital strategy, is embedded in education and can clearly demonstrate they're in it for the right reasons.

Co-production in action

Dogsthorpe Infant School in Peterborough (www.dogsthorpeinfants.co.uk) reaped the benefits of co-production, being instrumental in the development of NetSupport's primary classroom observation technology ReallySchool (www.ReallySchool.com).

> Observing and assessing each pupil every day to gauge their progress is a huge task. We tried several different solutions to try to make the process easier, but most of them only partially suited our needs, which caused more problems than they solved.
>
> When NetSupport approached us to test and feed back on ReallySchool, we were keen but also a little wary of the time it may take. However, we made sure to understand how the process would work, and our experience of co-production has been overwhelmingly positive.
>
> **Becky Waters**, Headteacher

> We knew that, in theory, ReallySchool would save us a lot of time – but it was great to test that out and prove it to ourselves. Knowing we've played a role in making the technology as useful as can be is really satisfying – and of course, we're glad to have had the benefit of early access. We can now tailor our teaching and learning so much more and be more active with planning, instead of simply rushing to complete our observations each day.
>
> **Rachel Jordan**, EYFS Learning and Teaching Manager

ESCHOOL NEWS MAGAZINE (ESCHOOLNEWS.COM) – MARCH 2021

3 big questions schools should ask EdTech developers

Partnerships between school districts and EdTech developers can be incredibly beneficial – but schools should make sure to ask the right questions at the outset.

Educators on Twitter know that sharing is something teachers love. If you're a school leader, it's likely your school is at the centre of your community, with close links to parents, local groups and valuable community initiatives. Working together is the whole ethos of a school. In the daily activities of teachers with their students, collaboration and teamwork is the heart.

Extending that cooperative spirit to embrace EdTech vendors is a beneficial mindset for schools. It's a great chance for teachers to provide real-world guidance and feedback to developers, helping them to shape solutions that fulfil a real purpose and with specific functionality the school needs.

I've worked in EdTech for nearly three decades and have held multiple voluntary roles at the school board level in the UK over the last 15 years, which gives me considerable insight into both sides of the equation. This unique perspective means I see the opportunities and obstacles for each party, but equally, can provide valuable insights for those schools and EdTech companies who do choose to work together. With that in mind, here are some questions to ask when considering the benefits of working with an EdTech vendor.

Do we have a clear idea of what our students and teachers need and why?

This question ties into your school's digital strategy, its aims, and whether being involved in a collaborative project with a vendor will achieve what is needed for your school or district. Without knowing your district or school's aims and where the impact is needed, time invested in a partnership will be wasted.

Is the company invested in the education sector?

Programmers not dedicated to education may not be a good fit. Conversely, well-established EdTech vendors understand the demands and pressures on teachers, administrators, and IT staff. I always recommend using a company that works primarily in education. Find out their commitments to other projects. Examine research they may have conducted and evidence gathered to prove efficacy. Is your district's primary need for technology like hardware, whiteboards, microphones, and tablets? Hardware solutions are far different from learning technology that must align to pedagogy. Have desired learning outcomes been factored into the equation? I would suggest that if the vendor's thinking does not align with your district's needs and values, you need to look for another vendor.

How much time is needed to build from scratch or modify an existing product?

Sometimes, people are hesitant to tread this path. I surmise that the problem is time. It may seem that the investment in time it would take to work with a company on a solution cannot possibly be worthwhile. But if both sides are clear on their objectives from the beginning, then being involved in co-producing an EdTech product can be productive and rewarding.

Firstly, find out from the developer whether they need input from scratch or if it is acceptable to test a completed product. Testing a completed product means that needed improvements and adjustments can be made from feedback, which saves time and money. Growing a product from scratch – for example, designing and developing a solution to record skills acquisition and progress for elementary school pupils, while integrating multiple reporting options and the ability to import data from third-party school management information systems – takes significantly more time than modifying an existing product.

Do not forget to factor in the long-term needs. Teachers are on the ground floor of implementation, so testing and collaborating can create extra work for them. Their feedback is critical not only to design the solution correctly at the beginning, but for buy-in later on. Investing time supports digital strategy and could lead to a free or reduced product price in the future – but only pursue this if the product fits your digital road map.

A final note on understanding the partnership

Finally, remember that transparency and frequent communication is key. Both sides need to respect the other's circumstances. Suppliers need to be clear about what they want to achieve with their product and set a realistic time frame for receiving feedback so that teachers have ample time to accommodate it into their lesson plans and workflow. Development timetables are tough to manage, so be sure to factor in any potential challenges so that production can continue uninterrupted.

Even with the best laid plans, student learning and security are the paramount concern, so software development needs contingencies. Yet, with a carefully chosen collaborator, co-production can have enormous benefits in bringing technology of value to schools. And if it can help you meet your digital strategy goals, it's a long path that has the best potential to produce a technology solution your school or district actually needs, rather than matching your school's problem to an existing solution.

PLANNING AHEAD

with your digital strategy

> Intuitively, I cannot escape a long-held belief that stability combined with the agility necessary to keep pace with global change, could enable our profession to become expert in its craft. To do this we need to create a long-term vision and plan for education which can empower a whole profession to join together in realising a nation's potential.
>
> **Carl Ward**, Chair,
> Foundation for Education Development (www.fed.education)

Whilst my main objective is to encourage questioning and reflection, the planning ahead and strategic stage is where we consolidate all the lessons learned, all the variables we need to consider, and find an effective way to deliver change. So, in many ways, this is the most important part of the book.

One topic I have often shared is how to shape a digital strategy for your school. Rather than provide the same narrative that I've already written and co-authored with Mark Anderson (@ICTevangelist) in our '**Creating a digital strategy for schools**' guide (which you can access from here – www.schooldigitalstrategy.com), we will explore the subject at a higher level and how the lessons learned can be adapted to suit your school's specific needs. To do that, I'll share some feedback and advice from trusted sources across the international education space.

A digital strategy (or any forward planning) shouldn't be centred on how much money is available. That might sound strange because clearly finances dictate what you can or can't have, but in far too many schools I've heard the same conversation play out repeatedly: 'We have £20k available – what can we buy?' We need to think about it differently.

Short-term budget spending is much like heading off on random short-term car journeys with no clear destination in mind. You can head off in any direction you like, and you may have some level of success, but a plan/digital strategy will help define your ultimate destination and ensure that each of those short-term trips/budgets takes you, step by step, in the right direction. The budget might dictate how many trips you will have to spread the journey over, but a digital strategy will act as the pin in the map and define the route to your ultimate destination.

As part of that planning, we also need a shift in how we think about EdTech within our schools – not simply finding where we can squeeze it in to add value, but being receptive to a change in the way things are done so that appropriate new technologies can be embraced.

Planning ahead will almost certainly become a catalyst for change. Change is often (but not always) a good thing and there are many variables that could influence the success of the process. Therefore, while we are navigating this chapter and reflecting on how we plan our digital change, it's also really helpful to be aware, at a high level, of the variables that might derail any proposed changes.

A useful model, often adapted to apply within the teaching and learning context, is the Lippitt-Knoster '**Model for managing change**'. The model is based around the five key elements needed for effective change – having a vision, gaining consensus, having incentives, allocating resources and having an action plan. If any element is absent or not fully engaged, then it will result in a variety of outcomes ranging from anxiety and confusion to resistance, frustration and so on. I've included an example visual below which will hopefully be a handy reminder as we discuss planning ahead in this chapter.

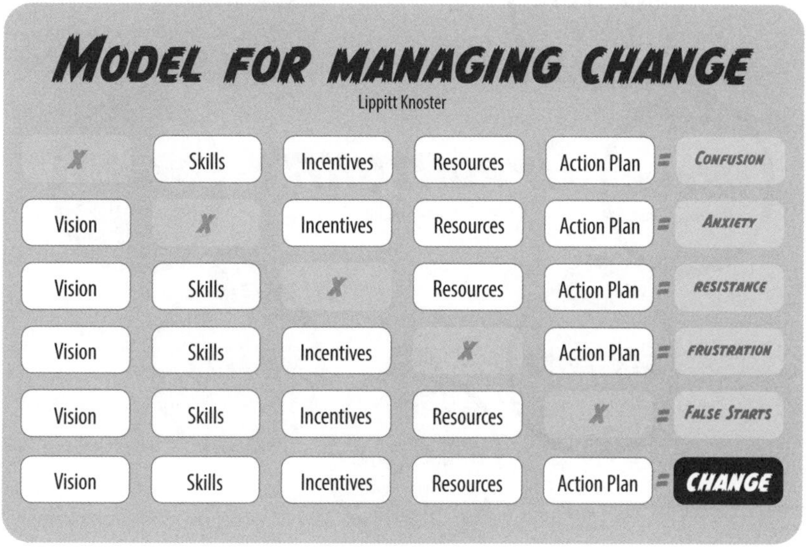

MODEL FOR MANAGING CHANGE
Lippitt Knoster
X Skills Incentives Resources Action Plan = CONFUSION
Vision X Incentives Resources Action Plan = ANXIETY
Vision Skills X Resources Action Plan = RESISTANCE
Vision Skills Incentives X Action Plan = FRUSTRATION
Vision Skills Incentives Resources X = FALSE STARTS
Vision Skills Incentives Resources Action Plan = CHANGE

Taking it a step further, it is worth considering the sequencing of events too, and Kotter's eight-step process for **Leading Change** (Kotter, 2012) encapsulates this well, highlighting that by encouraging a sense of urgency for change and maintaining that momentum, this process can be used to gain greater and, I guess, more timely impact in delivering change.

I think the image below is fairly self-explanatory and links to additional resources included at the back of the book, but in essence Kotter added the following to the change process model: creating a sense of urgency around change; building a consensus or coalition for managing the change; ensuring there is a communicated vision for the change; facilitating the removal of any obstacles to accomplishing change; promoting the continued pursuit of change despite any apparent success; and, of course, embedding the changes into the organisation's culture.

So now we have our gears fully engaged and we are thinking about how we shape our strategy, but we are also being mindful of how we might then need to deliver the resulting change and what potential barriers we may encounter.

From every challenge comes opportunity, and one of the clear benefits to come out of our accelerated exposure to new educational technology is that this experience has left us in a much better position to make informed decisions about what is needed next. In March 2021, I wrote an article for *District Administration* magazine (www.districtadministration.com) in the US and tried to condense the benefits of recent experiences, courtesy of COVID-19, into seven key areas:

1. **We know a lot more overall** about how technology functions in the real world in our schools than we did at the start of 2020. For many, that learning journey has been Odyssean, but we can now start to reflect on what we learned along the way and that's good.

2. **Remote/digital teaching** really helps some students and, whilst it has been a challenge and burden for many teachers, it has also been a catharsis and awakening for some too.

3. **We're all digital teachers now** and many schools have lots of trained and experienced digital educators. It may have been a sink-or-swim situation but the outcome – a veteran crew of tech-savvy teachers – is a definite asset.

4. **It's not theory anymore**, so while strategic planning is what it is, there are real, evidenced learning benefits that we can now use to support and grow. In other words, the planning and revision we must do now is no longer hypothetical or wishful – it's actual.

5. **There's no need for new hardware.** With the caveat that every school will be in a different place and on a different path, the chances are very high that revisions to your technology plan will not require a lot of expensive new hardware.

6. **We can make real progress without big cost** – the likelihood of not needing new hardware reduces implementation costs dramatically, increasing the probability of action significantly.

7. **Interventions and solutions can be targeted, even surgical** – knowing what parts of software created problems allows us to plan and target those things specifically. The ability to be precise can also lower cost and increase the speed with which solutions start to have impact.

It's always good to reflect on the positives, even if they don't all apply to all schools.

I was busy reviewing interesting posts on my Twitter feed recently and **Richard Spencer** (@Richspencer1979), an executive principal at **Cambridge Meridian Academies Trust** was sharing his top tips of what has worked best for him when planning improvements to teaching and learning. Some aspects succeeded more than others, but I really liked the points he made and they resonated well with readers, so I thought I would share the highlights. They set a good tone and have many parallels with the topics we will cover as we move into the specifics of planning our next digital footsteps.

1. **Prioritise expertise** – Reassure teachers that the development of their subject-expertise is your priority. Schedule time with their teams to work on this and improve their curriculum. Build networks with other schools to support this. Limit lonely actors.

2. **Establish focus** – Whole-school teaching and learning priorities remain an important collective driver. Review the evidence and form a working party to establish four/five key whole-school focus areas. Then strongly promote the implementation and evaluation of those strategies.

3. **Find your leaders** – Who are your examiners, who are your bloggers, who are your future leaders, who are your researchers, who is active in their subject association, who is ready for a new challenge, who will radiate the change you seek?

4. **Sample, don't scrutinise** – Learn about the student experience through appreciative inquiry. Review students' work and talk with them about what is working well. Do not audit books for compliance or expect rigid consistency – that way lies mistrust and a workload crisis.

5. **Eschew gimmickry** – The strategies that work will not be expensive or require days of consultancy to implement. Don't invest in outsourcing this work or expect a motivational speaker on an inset day to change the culture. Buy everyone books instead.

6. **Appreciative inquiry** – Seeking good practice will develop trust and help proliferate collaboration. Eliminate high-stakes observation. Normalise frequent low-stakes lesson visits. Do not allow appraisal policy to either dominate or limit your work on improving teaching.

7. **Consistency is a futile aim, but coherence is critical** – Most of us teach some weak lessons amidst very effective ones. All of us have bad days. Make it clear that there is no shame in failed endeavour at times. Promote radical candour within a culture of reflective sharing.

8. **Coaching can be key** – While highly effective, it is an investment and takes time. It also requires bravery and trust and can have unintended consequences. Leaders struggle with it because it is not a process they can easily track, audit or evaluate. Do it anyway.

9. **Don't offer inclusion lip service** – Make it everyone's responsibility to know more about and adapt more to the particular needs of learners with SEND. This applies at subject and whole-school level. Look for the barriers and build the bridges.

10. **Share, constantly** – Monthly bulletins, blogs, teach meets, collaboration, lab classrooms, inquiry projects, reading groups. Show your teachers: your agency is great, but so is the capacity to do harm. So stop, collaborate and listen to each other. Articulate sky-high standards.

I'm claiming ownership of the 'stop, collaborate and listen' quote (OK, Vanilla Ice might trump me on that claim, but I have a very close variant, as you'll see in the next section), but the rest of Richard's points are all salient advice.

Let's take the leap and head on into the next section in our future planning.

Stop, look and listen

In this section, we will be focusing on moving forward and defining our digital strategy, which we can hopefully construct around the key pillars: what we want to achieve; what we have learned; what has worked so far; and what we know doesn't work. As mentioned earlier in the chapter, I co-authored a digital strategy guide for schools with Mark Anderson and at time of writing this book, we are already on version 3 with close to a hundred pages of suggestions, best practice and peer case studies. At this point, I felt it would be useful to share the thought process and context behind some of the key topics and take slightly more of a helicopter view of the key strands.

If we try to compile the main drivers for undertaking a digital strategy (alongside the obvious COVID-19 influence), the most common responses are:

- Looking for ways to enhance learning outcomes
- Promoting student and staff digital wellbeing
- Increasing attainment
- Achieving better value for money (economies of scale)
- Developing collaborative technologies
- Fostering stakeholder engagement/communication
- Reviewing data security
- Reading Al's *Secret EdTech Diary* (okay, I made this one up)

In my presentations, I often start with a slide entitled 'Stop, Look and Listen' which really focuses our attention – reflecting on what we already have, where it is and if it's working well.

What springs to mind when you read the words 'digital strategy'? Is your first thought positive or negative? Or perhaps it doesn't really resonate with you at all? When I've conducted school inset sessions focused on digital strategy, I've often carried out a quick poll of the first words that spring to mind, using tools like **Mentimeter** (www.mentimeter.com). There are always lots of positive phrases like 'exciting', 'innovative', 'forward-thinking', 'future' and so on, but I also hear phrases like 'daunting', 'confusing' and 'capacity'. They are all valid and relevant phrases that we would expect to hear at the start of a discussion and journey.

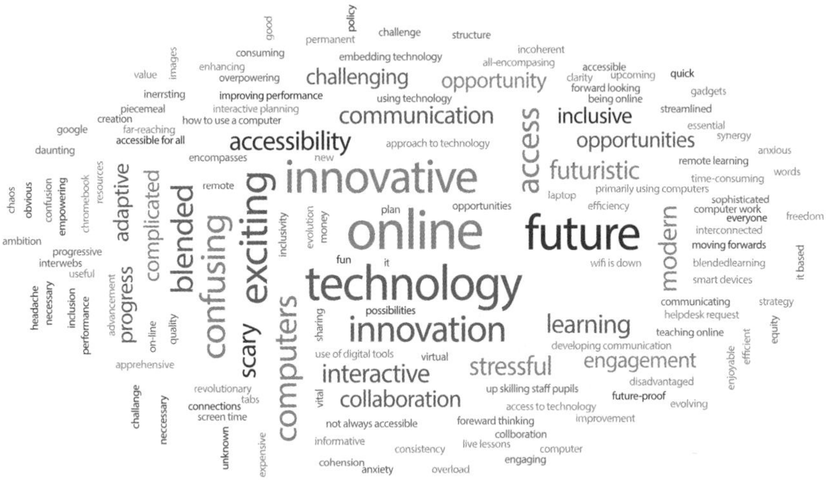

Example feedback captured during an inset session in 2020

Recognising and embracing those concerns, it helps to understand them within the context of something like the Kubler Ross '**Change Curve**' (Kubler-Ross, 1969) which describes the internal emotional journey that individuals typically experience when dealing with change and transition. The Change Curve journey consists of several stages that people go through – from shock and denial, anger, bargaining and depression through to acceptance and problem-solving. Whilst genuinely hoping to avoid any sense of anger or depression when reading my book, it is a useful and very relevant model to consider with some of the challenges faced whenever change is introduced.

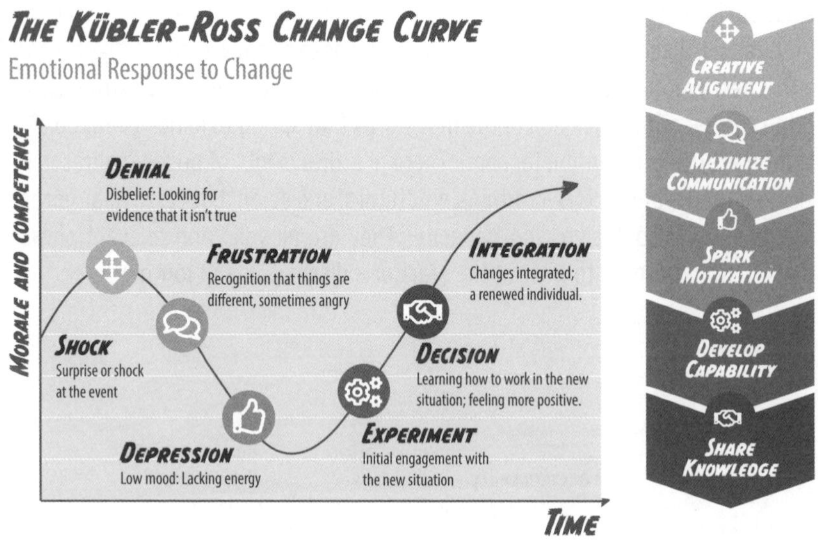

If we understand that everyone will have mixed emotions as we consider implementing a digital strategy, at this point we need to also recognise that we should never add to something we don't already have a clear grip or understanding of – strategies are rarely effective when they are simply adding more to something that doesn't work.

As another approach, some have chosen to build a digital strategy around the **ADKAR** model (www.prosci.com/adkar), which is one of the most widely requested and sought-after models for change management. Backed by 20 years of research, the model is based on the (often overlooked) reality that organizational change only happens when individuals change.

The five strands are:

- **Awareness** – steps to generate an awareness that change is needed.
- **Desire** – steps to generate a desire that change is positive for all.
- **Knowledge** – what information (and skills) needs to be communicated to facilitate change?

- **Ability** – what do you want people to be able to do with the information and skills?
- **Reflection** – how will the school and those within it be able to reflect on their progress?

Using this type of model can then feed into the broader road map and can be easier to align with any timelines that have been set for progress tracking.

Whatever method you choose, before you dive into the 'What do we want?' and 'Why do we want it?' conversation, consider a few key questions:

- Do we have a clear analysis of what has worked well so far – what tools and technologies have proven to be a good investment and have become embedded and used on a consistent basis?
- Do we know where our collective digital skills lie – have we undertaken regular skills surveys to identify where our strengths and weaknesses are?
- Are we clear on the priorities within our school's development plan so that these can be aligned with our digital strategy?

As part of that review, it's important to have a digital inventory to hand – a clear picture of what tech we've got (types of IT devices, their location, their frequency of use, which applications are installed) and our infrastructure and its ability to flex with any change. It's vital that you collect data first, then assess before you plan.

In terms of a digital inventory, often an IT review will discover underutilised devices that can be deployed elsewhere for better efficiency, or software subscriptions that are being renewed on an annual basis but never actually used, or devices that are perfectly upgradable rather than automatically replaceable – and of course all these factors can help with short-term cost savings.

By this point we should have a good sense of what tech we have, whether it's being well utilised and what works well, and we can then start to get a sense of what we would like to achieve. To refine that plan and ensure we have the broadest set of voices aligned (bearing in mind that it's much easier to implement change if we have significant buy-in), the next step is to bring the stakeholder voices and views to the table.

Using a stereotypical and exaggerated example, for many schools, the process of getting a decision on direction could be summarised as follows: business manager says there is some money available; someone asks for more of this item; item is bought and then the IT department are informed. There are so many opportunities for bad decisions in this process. The aim of course is to connect the dots – confirm that any solutions bought have the broadest possible impact and benefit, and that any decisions are made with an eye on the long-term aspirations – so that every investment has maximum value and is sustainable.

In terms of stakeholder voices, the first question needs to be: who gets a seat?

> **Teachers and students** – they can feed back what works well, what doesn't, what they'd like to be able to do with technology in the classroom, what examples they've seen at other schools, advice shared from peers on their PLN, what tools they find most engaging and so on.

> **Senior leadership team** – they will be able to review all the information shared, satisfy themselves that the priorities are being addressed, they are feeling consulted and informed so that they can make a judgement and sign off on any next steps.

> **Network manager and IT team** – they can articulate their own priorities for digital infrastructure investment and ensure that any proposed solutions from other stakeholders will not be limited by or negatively impact the current IT infrastructure; and, as with every stakeholder, it's always better to be informed and know in advance what potential changes are coming their way.

> **Data protection officer** – to ensure any new approaches and solutions do not compromise or negate the school's data protection and data privacy obligations.

Special educational needs and disabilities (SEND) team – to share their current challenges and aspirations and ensure, where possible, every child can access the chosen learning solutions, and where it is not possible, that dedicated solutions to meet each child's needs are factored into the roadmap.

Designated safeguarding lead (DSL) – again, sharing a view about any potential risks posed by e.g. developing our online resources or accessibility to technology, whilst also sharing any requirements they might have to further improve the school's eSafety solutions and promote digital citizenship.

CPD lead – to ensure that whenever the addition of new technology is discussed, the investment of time and budgets for suitable training is factored in to deliver the best possible chance of the technology becoming embedded, being used regularly, and having a tangible impact.

School business manager or trust finance director – to listen and understand the needs being articulated and, once prioritised, identify the funds available and the pace that will be needed to move towards the ultimate goal. In other words, the money shouldn't dictate what happens, just how quickly it can happen.

Trustees or school governors – for oversight of any investment and to ensure constructive challenge of the conclusions reached.

The above is a simplistic summary of the voices to involve and there will no doubt be others we can include, but it should be a good starting point to at least set the aspirations. As those ideas and requests are distilled down, we must also ensure that we remain focused on our overarching priorities, which should be shaped around pedagogy, alignment with the school development plan, being realistic within available capacity, having the best potential to be sustainable and reflecting on ways to measure and evidence impact. Or, as our good friends at Ofsted would say, we must be clear on our **intentions**, our plan for **implementation** and an agreed approach to measure and ensure **impact**.

What we are hoping to achieve can be encapsulated within these statements:

- A clear vision of what we are trying to achieve.
- An action plan of when, where, and how different things will be undertaken.
- A strategy to ensure staff are aware of who will be expected to use these technologies.
- An infrastructure to support these efforts.
- Training to ensure everyone knows how and when best to use technology and understands how it can support learning.

At this point we are formulating our vision rather than the actual strategy. As **Linda Parsons** (@DigLin_) at Deira International School explained: 'Don't start writing a digital strategy until you have tailored your school's digital vision! The vision you are about to write will stand as your guiding light for future applications of technology in your school, and also provide the solid foundation for all starting points of digital conversations.'

The intention of this book has always been to merely bring together all the points for consideration and to spark further discussion in school, not to act as a revelation to most readers. However, it is worth noting that whilst I have shared a simple 'who's at the table' summary, in practice it is a much more nuanced approach to ensure the right discussions are undertaken. Most MATs, districts or schools will have their own process to bring the key strands together and many, like ours, will choose to break down the discussion points into strands (or – in our case – pillars).

Breaking the discussion down into six key pillars has proven to be an effective way of keeping the discussion on task for each group. Each group consists of not just the lead practitioner of a topic, but a range of staff who have digital skills or experiences from other schools, so we get the widest possible range of voices in the discussion.

Our pillars were framed around:

- Innovating learning
- Developing students' digital skills
- Developing teachers' digital skills
- Technology and infrastructure
- Communication (interwoven across the other sections)
- Wellbeing

This process was initially shaped by looking at other schools' examples, so it is by no means unique or intended as an exemplar for all to follow.

Innovating learning

Everyone will have their own views on this, and in truth, that's kind of the point – innovation isn't a fixed formula that, once prescribed, delivers guaranteed impact, and as you'll read in the next section, ensuring the choices made are well-informed is key.

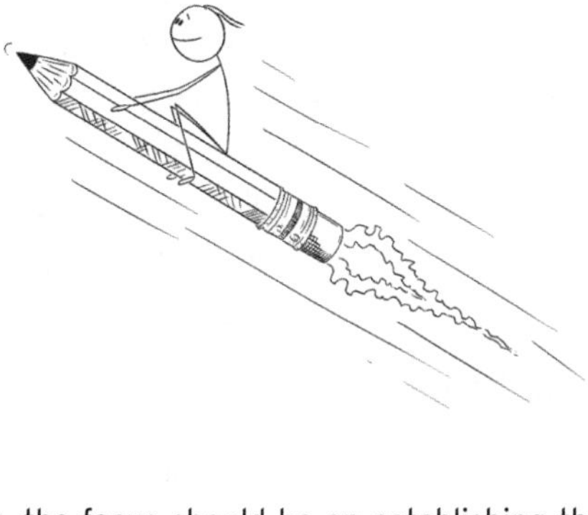

To innovate learning, the focus should be on establishing the current use of ICT across the trust and identifying the most appropriate tech for the subjects taught. It's very easy to be convinced, for simplicity and consistency, that the pathway of a single device is the better option, but there are strengths and weaknesses to all offerings. Key stage 5 business studies students want access to devices that are consistent with those used in the workplace (i.e. by their potential future employers) for information research and writing; KS3 science students might want tablets that support add-on measurement tools to record experiments or augmented reality apps to help them visualise the dissection of an eye or heart. Clearly some of these decisions are shaped by the platform, others by the availability and richness of the app ecosystem. A simple SWOT analysis (strengths, weaknesses, opportunities, threats) or something similar can help identify the best device aspirations for each department.

Alongside the review of devices and apps technology, reflecting on the needs of our SEND/ALN learners needs to be part of the planning, looking at supporting tools such as immersive readers, to help mitigate any issues.

Given recent challenges, some of the discussions within this pillar centred on the right blend of remote learning (the balance between synchronous and asynchronous), the effective delivery of feedback to students and ways to measure student access and engagement with the resources provided.

Once technology needs have been identified, discussions should focus on how best to utilise and integrate existing and new EdTech into an enhanced curriculum delivery. Approaches like those used in **design thinking** (Arnold, 2016) can be developed alongside promoting the use of suitable apps that foster independent thinking, problem solving, critical analysis and so on. There is a wealth of apps available within the core (Microsoft/Google) desktop suites that can support prototyping and testing within projects (think Forms, Adobe, Spark etc.).

Developing students' digital skills

Many of the elements in this category have already been covered within the earlier 'Lessons Learned' chapter, but the key strands were about ensuring we had an accurate baseline of our childrens' digital skills and putting in place suitable training and interventions to develop and build confidence in the use of core apps. But be clear on expectations – digitising resources, for example, won't deliver direct teaching and learning gains but it's reasonable to assume that efficiencies and cost savings can be leveraged.

It would be easy to assume that our generation of digital natives all have the skills required to access any technology placed in front of them but in practice, that simply isn't the case. Building confidence levels (including in their parents, where appropriate) is key, as is the need to manage and shape expectations – from how to interact and behave in a remote lesson, to getting the most out of using Microsoft Teams, to being comfortable accessing, completing and returning assignments, for example. This isn't a one-time process. It must be a continuous journey to ensure confidence levels remain high, and I believe all schools will recognise the need to revisit this regularly and find creative ways to ensure levels are maintained. Ongoing skills surveys are a simple and effective way to measure the impact of a school's coordinated efforts.

We must also recognise that more can be done to develop our students' digital citizenship skills too. We know that the greater the access to online resources, the greater the potential to expose a child to risk, so being proactive about the tools and strategies adopted is fundamental and vital and can be condensed into three key areas:

- **Content** – being exposed to illegal, inappropriate or harmful material.
- **Contact** – being subjected to harmful online interaction with other users.
- **Conduct** – personal online behaviour that increases the likelihood of, or causes, harm.

That means reviewing existing policies and procedures to ensure compliance with statutory obligations, making sure appropriate monitoring and filtering solutions are in place (as well as ensuring any new apps or devices introduced as part of the broader process don't create any additional vulnerabilities), and regularly undertaking a data protection impact assessment (or equivalent).

I will also repeat here that children must have a clear route to proactively report a concern, ideally being able to select a trusted member of staff to bring any issues to. More than ever before, this is a really important consideration, as is having the right tools to log, track and (if appropriate) engage with external agencies using key tools like **CPOMS** (www.cpoms.co.uk) or **MyConcern** (www.myconcern.education).

The most powerful long-term solution though, as mentioned previously, is to teach our children to be good digital citizens. We need to shape the way they see and engage with the online world, how they communicate and, crucially, how they assess and validate the accuracy and authenticity of what they read. Promoting awareness and understanding of data privacy, digital footprints and the importance of both minimising what we share and being careful who we share it with runs parallel to this.

In summary, the focus for our students in terms of technology should be on identifying opportunities to develop creativity, how it can improve communication and collaboration, improved digital accessibility, additional safeguarding measures and, of course, time efficiencies.

Developing teachers' digital skills

As with students, the primary focus should be to understand the current digital skillset of all teaching staff (including teaching assistants) – to identify not only areas for development but also where the pockets of strength and experience lie. One of the best strategies to support staff confidence is to introduce 'flag-bearers' within the school – the 'go to' staff who can support others on a particular platform or application. This approach is not only good for staff but also healthy for the school as it eliminates the risk of one person becoming the expert and the inevitable consequences if they move to another school.

CPD is something I have mentioned several times, and the lack of focus given to this area, particularly around technology, is a common theme I've seen in schools (and business) for decades – buy new tech, subscribe to new apps, spend all the budget and then expect staff to magically figure out how to use it and the IT team to know how to effectively support it. It's a recipe for failure and it happens far too often to be a coincidence. As mentioned previously, vendors also have a role to play in providing more accessible solutions and better product-specific CPD for customers.

Alongside CPD (and not just a day of inset, but a rolling programme of training), where there are opportunities to consolidate key platforms into a common one across the school, trust or district, then of course that makes it much easier to embed and share resources. As well as the tech flag-bearers, some schools (particularly MATs) are scheduling regular tech clubs (in a TeachMeet format) to share ideas, new apps discovered, top tips etc. We held one recently and it was well supported, with all the ideas broken down into short five- to ten-minute demonstrations.

Although classroom AV (audiovisual equipment) is wrapped up within the broader process, it's worth mentioning here as it often surfaces quickly – the tools teachers want to use and their confidence in getting the most out of it. Although many teaching staff have interactive whiteboards, the majority simply use it as a surface to write or project onto. Some have utilised resources like digitisers to allow for easy projection and, as we move to a more mobile delivery within the classroom, switching the IWB to a TV with suitable device projection tools is a better strategy that unlocks fresh opportunities within the classroom. The most common feedback is that classroom AV untethers teachers to be more involved in the classroom – that alone justifies highlighting it here.

To close the loop, the final checkpoint was to initiate regular surveys to capture feedback and progress and expect the requirements to change over time.

Technology and infrastructure

I've talked a lot about technology and infrastructure, and as you would expect, the topics raised feed into this pillar of our strategy. Once the nuts and bolts of the infrastructure are in place andthere is confidence in its capacity (connectivity, storage, access) and ability to expand, this strand focuses on how the infrastructure can help develop communication, tools and connectivity to support and build parental engagement – from school to home and the broader community, utilising the infrastructure to deliver better access to training resources (for example, developing the use of Microsoft Stream for training resources and asynchronous learning resources whilst also ensuring strong data protection policies are in place, staff are GDPR-aware and so on).

Much of the above builds on tools to manage the IT infrastructure and be more proactive in terms of alerting when issues might occur, facilitating safeguarding requirements from a digital perspective and looking at ways to standardise platforms where possible. In a MAT, when new schools arrive it's very easy to end up with a mixed Microsoft and Google IT estate. This sometimes happens at department level in the absence of a joined-up strategy and not only creates significant management overhead for the IT team but also requires multiple logins for staff and students and potentially doubles some of the CPD required. Consolidation is therefore an important consideration as part of the review.

Wellbeing

Wellbeing should be a separate pillar within a good digital strategy, but in truth it's interwoven within all the other pillars:

- Identifying the core platforms and tools to support and foster better staff communication (professional and personal).
- For the internal PLN, utilising those core tools to help reduce teacher workload (whether through easier scheduling, auto tracking activity, auto marking via rubrics, making it easier to provide feedback, perhaps with voice notes etc.).
- Post-pandemic, reflecting on how the blend of classroom and remote working impacted on wellbeing, expectations of staff and a realistic workload capacity.

- Having a staff wellbeing group, as well as access to employee assistance programmes for support and advice relating to any issues in their professional and personal lives.
- Fostering a culture where we make time for the small things that have an impact – like taking the time to stop and simply ask, 'How are you?' Not sending emails after 6 p.m. is another helpful and easy-to-apply approach.

For our children, the pandemic has hugely elevated the focus on their social, emotional and mental health. There are many strategies that are really important for students' emotional development: enabling easy access to support guides and resources for learners; being accessible virtually face-to-face, especially for our most vulnerable learners; signposting and scheduling online activities in advance to provide routine and structure; and using EdTech to help foster student-to-student peer communication and collaboration.

Within the fabric of the school, it is important to utilise digital signage to promote wellbeing, celebrate student success, showcase the school's visions and values, and embed that into lessons.

I'm sure you'll agree, you can never do too much on this topic, and there is a constant stream of ideas available from others who have had success and impact.

Different approaches to consider

I've seen some great examples of schools embracing digital change that are definitely worthy of note, like the work undertaken by Linda Parsons (@DigiLin_) and Paul Gardner (@DubaiDeputy) and the team at the Deira International School (www.disdubai.ae). They won the ISC's 2021 Digital Technology in Learning award at the International School Awards, and I've included a summary below from Simon O'Connor, director of DIS, who explained:

> The use of digital technology is at the forefront of education development, and this highlights that Deira International School is leading the way and helping to construct the future for all our students.
>
> Teachers at Deira International School have been working hard over the past 18 months towards a common goal. As part of our now award-winning digital strategy, teachers have embraced the medium of technology to enhance the already outstanding teaching that takes place. The purpose of technology in education is not simply to replace what has previously been done. Instead, technology gives us the opportunity to redefine how learning takes place. Students at Deira International School are direct beneficiaries of this digital strategy. Every day at school, students benefit from the high level of aptitude from teachers with how to use digital platforms to create the optimal learning experience for students. It is important to note that this award reflects the quality of teaching and learning through our digital provision, not our digital provision itself. As such, parents enrolling their children at Deira International School can be assured of the highest quality education.
>
> The use of artificial intelligence is now integrated into our school assessments, particularly in English, maths, science and Arabic. Teacher feedback to students is often via verbal voice notes through Microsoft OneNote. The use of verbal feedback adds significant value to the feedback process. The ability to give more depth through the use of voice notes is a game-changer when it comes to facilitating student progress. Our use of Microsoft Teams, in particular the assignments and rubrics, means that students have objective and meaningful

> feedback based on commonly shared criteria. Part of our digital strategy includes the readiness to adapt to the ever-changing opportunities that are available to us. This means that whilst we have various integrated apps and processes, we are always looking at how to make this even better.
>
> A key part of our digital strategy is the leadership of our student body. Students at Deira International School have been a fundamental part of developing our vision, creating our strategy and implementing it. Students have learned about the leadership process and how to engineer change. This is being embedded at all levels within the school.

They have a five-year digital strategy with five strands (like my pillars), broken down into digital communication, staff CPD, device strategy, digital citizenship and educational technology. It's a great piece of work with the building blocks of developing communication to deep integration of key learning platforms, through to crystallization of skills and abilities over time.

Olly Lewis (www.ollylewislearning.com), from the British International School of Dubai, explained their approach:

> We developed our digital strategy by firstly exploring our curriculum to see areas where technology:
>
> - Was already being used well.
> - Needed improvement.
> - Could provide opportunities for development.
>
> By doing this and then closely evaluating the technology we already had in the school, plus running a full digital audit of the hardware, software and digital subscriptions we already had, we were informed of our strengths, weaknesses, opportunities and threats (SWOT analysis). This gave us a clear roadmap of the areas we could capitalise on in terms of teaching, learning and our infrastructure and where any potential cost savings could be made.

> Our digital strategy was constructed around the following principles:
>
> - Streamline effectiveness and efficiency.
> - Improve accessibility for all stakeholders.
> - Foster creative teaching, learning and professional development opportunities.
> - Secure data and safeguarding within our ecosystem.
>
> The most prominently used tools to facilitate teaching and learning are Microsoft Teams, OneNote Class Notebook and Microsoft Stream. As our approach is a blend of both synchronous and asynchronous learning, with an emphasis on accessibility and inclusivity, we provide learning opportunities for students which are equitable, have suitable academic rigour and promote collaboration.
>
> We decided that for our digital strategy, the best method for teaching and learning with technology was for each cohort to have a simple and uniform approach, regardless of year group within the secondary school. We aimed to provide consistency in the usage of Microsoft Teams and OneNote Class Notebook and ensure that, included within each, there was a channel for each subject.
>
> Having this consistency across the board for our learners and teachers makes navigation, communication, accessibility, planning, sharing and knowledge of the tools as easy as possible for all stakeholders. This strategy makes everything from the subject channels to the wellbeing channel easy for students to navigate through, helping to drive efficiency and effectiveness.

There are plenty of other great approaches available to review and many schools share theirs in the public domain. If you jump into Twitter and engage with the community, you'll find plenty of sage advice and examples you can draw from. No one approach is right – it's about what's right for *your* school – and it's about taking small steps.

Asking the right questions

Now this is an interesting and challenging topic to round this chapter off, and potentially slightly contentious.

Let's start by acknowledging (and hopefully agreeing) that whatever we choose to add into the mix in our school's digital strategy 'needs to be fit for purpose and suitable for our needs and, ideally, have a proven track record'.

At a simple level – should you buy tech based on the website description and supporting brochure? The answer is, of course, no. We need to consider how to move from cautious (and possibly pessimistic) to reassured and optimistic. In other words, we need to try to avoid digging a hole for ourselves.

Let's break down the journey from cautious to reassured:

> **Try before you buy** – nothing is more reassuring and convincing than trying the tool within your own environment, on your own equipment and with your own students to develop confidence it has value. So, always request a free evaluation copy of the solution first.

Review case studies – for brand new solutions it might be tough, but for more established solutions, look for case studies from peers who have used the solution. Ask your PLN for advice – seek external validation where you can. It makes sense to do this before spending too much time on the evaluation stage, especially as with some solutions (e.g. offering curriculum resources or perhaps retrieval tools) it may not be that easy to evidence during a short evaluation cycle (so it never hurts to ask for an extended evaluation if you can't get feedback from peers).

External validation – if the solution you are considering is pedagogy aligned, look for external validation of the product's focus or strengths. A good example would be something like the certifications from **Education Alliance Finland** (www.educationalliancefinland.com), who provide product evaluations and certifications based on common standards for learning solutions which are undertaken by Finnish teachers and researchers. Their feedback helps shape and improve a product's pedagogy and is also a useful and respected way to validate the educational benefits of a solution. As a fan of co-production and pedagogy-based solutions, I give them a big thumbs-up and am proud to have solutions rated in the 90%+ range by them. As I always say, you can't talk the talk if you aren't willing to walk the walk!

As Olli Vallo, CEO of Education Alliance Finland, explained:

> We specialize in the pedagogical impact evaluations of learning solutions globally. In the EAF evaluation process, the learning solution is mapped against learning science principles, and its curriculum alignment and usability are assessed. Through EAF evaluation, EdTech solution developers will get rigorous feedback on the strengths and development areas of their solution. The impact evaluation report and certification provide evidence on the

solution's educational quality and value. EAF evaluations are done by a trained group of teacher-evaluators, using evaluation criteria developed by educational psychologists from the University of Helsinki.

You would probably be thinking along the same lines when evaluating any solution yourself, but there has been a strong move towards choosing research-informed solutions and ensuring that schools make the right decisions. The **EDUCATE programme** – led by UCL's Institute of Education, in partnership with Nesta (www.nesta.org.uk), Besa (www.besa.org) and F6S (www.F6S.com) – is helping EdTech start-ups to use existing evidence and collect new evidence as they test and adapt their products.

This shifting narrative is to help vendors understand the expectations of the solutions they provide, which can be distilled down into five key points:

- Vendors should have a clear understanding of where their product fits in within the school.
- Vendors should understand, and be able to evidence, any claims that their product creates impact.
- Where appropriate, vendors should be using available evidence and research to design their products.
- Vendors should be responsive to feedback from, and co-production with, educators and students.
- Vendors should not overstate their product's impact.

In the UK we have active groups like the EEG – **the EdTech Evidence Group** (www.edtechevidencegroup.com) – who provide checklists and guides on what good evidence looks like. Although I think it's a good concept and well-intentioned, it remains the case that, like every child and every school, no two products are alike and the breadth of EdTech solutions don't fit neatly into the same boxes. Compare an EdTech IT management app with a maths curriculum app, and whilst there will be some commonalities (in terms of rationale and user feedback), there aren't many. It's therefore important to keep a broader set of measures in mind, based on what you're specifically looking for.

In March 2021, Dan Sandu wrote an article entitled '**Show me the evidence: How do you procure your school's EdTech?**' (www.sec-ed.co.uk) where he included advice during the evaluation process, and he advised that, given the short evaluation window, schools should plan ahead and during the trial period:

- Get advice from the company and tap into their expertise.
- Recruit a group of teachers to try the product and agree together beforehand what you want to test.
- Review the trial regularly as a group to gather as much information as possible.
- Get feedback from students and any other stakeholders that have been involved.
- Try to gather quantitative and qualitative data if possible.

In April 2021, Brian Seymour (@SeymourEducate), director of instructional technology for the Pickerington Local School District in Ohio, wrote a great summary for EdSurge (www.edsurge.com) entitled '**Trust the Process: How to Choose and Use EdTech That Actually Works**', where he expanded on their digital content evaluation and selection process. As well as providing a detailed checklist to follow when selecting new digital solutions, he highlighted the different types of evidence to consider as part of the process and (rightly) pointed out that context matters – what works for one school might not work for another – and therefore suggested schools consider the following types of evidence during their evaluation and decision-making process:

> **Anecdotal evidence** – impressions from users' experiences. While this type of evidence is the easiest to obtain, it's also the weakest form of evidence because it's based solely on an individual's impressions. Common sources include blog posts, testimonials, promotional videos, recommendations and reflections.
>
> **Descriptive evidence** – measures of outcome over time. This type of evidence provides basic descriptions on potential impact, which is common and easy to find in marketing materials and news articles, but leaves out information about critical factors that may have influenced the outcome

(e.g. teacher, classroom and curriculum factors). Common sources include white papers and pre/post examination summaries.

Correlational evidence – comparisons of users and non-users. This type of evidence identifies a relationship between the use of a solution and the non-use of a solution but doesn't demonstrate direct causality and cannot be used as a conclusive result. Common sources are white papers, comparison charts and independent research reports.

Causal evidence – accurate measures of effectiveness. This type of evidence limits the solution as a single variable, hence it is the only reliable method for demonstrating true effectiveness, but this type of evidence is difficult and expensive to conduct. Common sources include research journals, summaries or peer-reviewed articles, and independent research reports.

I've included a link to his evaluation and selection process checklist (Seymour, 2021) within the 'References' section of the book and I would really encourage you to check it out.

To add another perspective for consideration, I read a great article called '**Meta-EdTech**' in *Learning, Media and Technology* by Ben Williamson (2021). In it, he highlighted that with the rapid rise and interest in EdTech solutions, two new types of organisations and technology platforms became more evident during the pandemic: firstly, EdTech 'evidence intermediaries' designed to educate leaders, teachers and procurement specialists about the evidence for 'what works' in educational technology; and secondly, EdTech market intelligence companies that help investors 'learn' about the value of investing in educational technologies.

He groups these together under a new term: 'meta-EdTech' – emerging EdTech solutions built and hosted by intermediary organisations which operate between different constituent groups within the EdTech ecosystem, such as **LearnED** (www.LearnED.org.uk) or **EdTech Impact** (www.EdTechImpact.com) (see below). He also concluded that these types of portals will become increasingly more influential in the coming years.

Although this topic is important for schools looking at new solutions, it has equal importance for new start-ups, as well as established vendors, to maintain a 'finger on the pulse' of the EdTech landscape.

EdTech 'evidence intermediary'

The EdTech 'impact and evidence' specialists and networks belong to an emerging group which work by compiling evidence about solutions for schools. For example, here in the UK we have sites like **EdTech Impact** (www.EdTechImpact.com), an independent review platform for education technology. For an annual subscription, EdTech companies can list their products, receive 'verified reviews' and earn 'teacher choice badges' to help promote their brand.

In the US, the **EdTech Evidence Exchange** (www.EdTechEvidence.org) is a partner of the 'Learning Keeps Going' coalition of EdTech industry providers and supporters convened by the International Society for Technology in Education (ISTE) to support learning during COVID-19 lockdowns. Its evidence framework is known as the EdTech Genome Project, a large-scale collaboration between education technology researchers, the EdTech industry, educators, entrepreneurs and advocates, and was launched in 2019.

As Williamson concluded, these evidence and impact groups act as new kinds of intermediaries within education systems, brokering consensus regarding the implementation of EdTech through the creation and presentation of novel forms of evidence and product credentialing systems.

Although I wholeheartedly support the best endeavours and intentions of these groups, I would also like to flag a small fear. It's good to have standards to measure against, and it's good to define different ways vendors can evidence the credibility of their solutions, but as we head down a path where potentially a few people become the de facto gatekeepers of the sector, and the only way to join the party is by paying to be included on their portal, I worry that we risk losing the level playing field or, more importantly, the independence of the sector. I have always felt that the strength of our sector is in the breadth of voices and the innovation they represent. Every time we start a process of distillation, I have to question whether hearing from just a few voices truly adds value. There is strength in unity, but like most things, there are pros and cons. It's just something for us all to reflect on.

EdTech market intelligence companies

For completeness, the second group defined are the '**market makers**'. These are the market intelligence agencies who assess the market value of educational technology companies and support venture capitalists to invest in products with high predicted future value. I'm not sure these are as relevant in the context of this section, but they have the taxonomy of the market, the value predictors and an eye on the growth areas. To put in context, one such market maker, HolonIQ (www.holoniq.com) calculated that in EdTech, venture capital investments alone globally exceeded $16 billion in 2020 – up 320% since 2010 – and EdTech is predicted to continue to be a key growth sector moving forward.

In summary, the good news is that there will be an increasing source of information and resources to ensure schools find the right solutions, and – as they like to say on a popular quiz show – if in doubt, 'phone a friend' (or a peer at least!).

SECED MAGAZINE – SEPTEMBER 2020

Planning your school's digital strategy post-Covid

With the future still uncertain, digital plans for schools have taken on a new significance and we must ensure that EdTech can be used effectively both in school and remotely. Al Kingsley offers some advice and reflections.

Whether you are starting a digital strategy from scratch or simply adapting it, the first thing to do is reflect. As well as identifying the areas that are digital priorities, it is also necessary to look backwards to get a clear picture of the technology currently used in your school – and if or how that has changed during the lockdown. You can then decide which solutions are effective and are delivering impact.

The word 'impact' itself can sometimes be a barrier as it may give the impression that everything must deliver measurable evidence of progress. However, it can be more than that. It is about saving time. It is about saving resources. It is about promoting wellbeing – and much more. Some of those things are less tangible when it comes to measurement.

- For example, recent months have shown that using tools like Teams, Hangouts and others has significantly helped with peer-to-peer and teacher-to-teacher engagement and collaboration. Those kinds of benefits are not things that necessarily filter through to school data and results.

At the heart of your digital strategy are students and teachers. The core areas to consider are:

- Enhancing learning outcomes and supporting pedagogy.
- Increasing staff, student and parental engagement.
- Allocating training time to ensure teachers are confident with using the tools (especially important when thinking about trust-based operations, where staff are potentially required to work in different locations).
- Implementing collaborative technologies.
- Thinking about how, as a school or trust, technology can be used to promote digital wellbeing.
- Employing sustainable, cost-effective solutions.

It is worth noting too that a clear digital strategy can deliver additional benefits for a MAT or federation of schools. For instance, there are significant economies of scale when it comes to buying technology collectively in bulk, rather than piecemeal as standalone schools. Standardising solutions across all schools in a group, as well as centralising their control and maintenance, can also help achieve better value for money.

Three golden rules

- **Be clear** – First and foremost, keep it simple. Complicated strategies (and/or revisions) are often less flexible and more likely to disenfranchise the whole school community. It is much easier to concentrate on one or two key changes and ensure sufficient time for CPD to build staff confidence than it is to try to introduce lots of changes at the same time.
- **Recognise when tech is needed – and when it is not** – It is important not to fall into the trap of using technology for technology's sake. The question to ask that gives you maximum insight into your school's IT situation is: would anyone notice if it was gone? It is important to recognise that technology is not the panacea for everything; it is simply there to support good teachers in delivering great lessons.
- **Work within your budget** – Start by looking at where your existing technology can multi-task and bring you savings (in time, money or both) and also at technologies that you are paying to lease or maintain, but which you are not really using. You can then redirect the money saved to a different area of your strategy. Ensuring your plan is sustainable over time and that your existing technology will continue to add value will provide consistency for everyone.

EdTech post-COVID

How has COVID-19 affected how we think of a school digital strategy? We all know that the method of delivering teaching and learning has changed fundamentally in the last few months and, for many, there is no going back. There is now a new emphasis: the requirement for schools to consider what technology will work best for them both inside and outside of the classroom.

Blended learning

The biggest change is undoubtedly the use of the blended learning model, and I believe that this is here to stay. I have heard countless stories of its benefits, particularly regarding engagement, and it offers greater flexibility for both teachers and students. For example, it could be used on snow days so that students do not miss a day of learning, or for delivery of revision classes during the Easter holidays, so students and teachers do not have to come into school.

Of course, more technology-based remote learning throws up its own challenges – namely, the digital divide. Technology itself cannot fix the challenge of students either having no access to it at home or access that is limited or shared with parents and siblings. So, with that in mind, blended learning is likely to be best employed as a supportive platform alongside more traditional methods.

Other tech possibilities

At the core of every digital strategy is the need to make evidence-based choices about classroom technology that supports pedagogy. However, we also need to consider whole-school technology, particularly the role it can play in student safeguarding and pastoral care, on and off site.

The ability for teachers to use technology to maintain one-to-one relationships with their students – whether in the classroom or over a remote connection – will remain vital in the months to come, especially for quiet or vulnerable children. It will be important in order to build normality into their schedules as well as to provide reassurance where it is needed.

Now that the blended model has come into play, another thing to think about is ensuring that teachers have the tools to create and distribute resources effectively – and that students can return their work just as easily. Choosing the right technology can really streamline this process and prevent an unnecessary extra burden for teachers.

In addition, EdTech that enables teachers to provide timely feedback to students about their progress is critical to maintaining learning momentum and motivation (especially for remote learners) – and tools that can help provide enrichment activities also need to be on the list for consideration.

Reflect and revise

For many schools, COVID-19 has been a catalyst to start the conversation about digital strategy. Some were already some way down the path; others less so, and they have had to catch up quickly. What needs to happen now is that the areas that have benefited from an online approach through necessity (e.g.

collaboration, communication and pedagogy) are not lost as 'normality' returns. There are ways that many of those facets can be incorporated into the standard methods of delivering teaching and learning in classrooms.

So, when it comes to digital strategy post-COVID, be clear on what you want to achieve. It doesn't have to involve spending lots of money; it's about focusing on what works well for you in your context, while ensuring that teachers and students have the confidence to use tools effectively – creating a springboard to creativity and innovation within your EdTech environment.

Find out more about creating a school digital strategy at www.schooldigitalstrategy.com

Voices Aligned

In a sector with so many people now advocating the need for change and the need to embrace the opportunities EdTech presents, I wanted to share the thoughts of a few trusted peers to help develop and amplify the message. I asked a selection of key questions and aligned their responses with each question.

My sincere thanks to all who took time to share their thoughts with me. I hope their views and aspirations resonate with you. We are genuinely blessed to work in a sector where people are so willing to take the time to support and share.

How should we consider EdTech and its future role in education?

Jon Tait

'Technology should be used in innovative ways to support and enhance learning, where digital tools make things more accessible, engaging, efficient and intelligent.

However, we should not just be duped into thinking that just because we can digitise something, or a process, that we should do. Many forms of analogue methods are still the best and digitising them just creates more cognitive load for the end user.

It's about finding ways where digital technology can enhance an analogue process and not just replace it because it's shiny.'

Jon Tait – Director of School Improvement and Deputy CEO, Areté Learning Trust
@TeamTait

Gary Henderson

'The world we live in increasingly relies on technology, and therefore technology should also, rightly, increasingly be a part of education. In addition, technology can be used to better engage students and can provide students the flexibility to build and demonstrate their learning in a way that suits their needs and preferences.

I also prefer to refer to 'technology in education', as I sense it conveys the broader use of technology in schools, colleges, etc., than 'EdTech', which to me conjures up views of IWBs, VLEs, etc.

Given the pandemic, it is clear we need to have solutions, including technology, in place to deal with future situations which may require school closures, ranging from another pandemic to something as benign as a school snow day.'

Gary Henderson – Director of IT
@Gary Henderson18

Philippa Wraithmell

'Reducing workload, helping ALL learners thrive, making education globally collaborative, connected, and exciting. Bringing the real world to the classroom.

The best support for every child, supporting, stretching, challenging, making sure no child gets left behind, no assumptions of understanding are made and most of all every child has the opportunity to thrive being supported in their learning in a way which one teacher in a room cannot. The support of EdTech in the classroom allows for every child to be their best self and to be supported when needed.'

Philippa Wraithmell – Director of Digital Education & Innovation
@MrsWraithmell

Sophie Bailey

'The role of EdTech in education is important in order to equip learners with digital skills to navigate the world around them but also so that they may become self-directed, connected learners throughout their lifetime, able to become more knowledgeable about their passion projects as well as skills for work.

EdTech should improve their quality and enjoyment of life.'

Sophie Bailey – Founder and Host, The EdTech Podcast
@podcastedtech

Scott Hayden

'I want us to think about EdTech as we do a whiteboard, post-its, and pens – as tools. The sooner we stop thinking of EdTech as 'other' e.g. online/remote/digital etc. the better.

Learners do not think 'I'm doing online and digital learning now' – they are just learning. To not prepare learners for how to use technology is a dereliction of our duty as educators to prepare young people for the future.'

Scott Hayden – Teacher and Digital Innovation Specialist
@ScottDHayden

Caroline Keep

'EdTech needs to have a bigger role in education. All technology does for education! We need to move away from just devices to electronic hardware, 3D printing, physical computing, IoT and machine learning.

We have only just felt the slightest impact of the fourth industrial revolution. If 2020 taught us anything, it is that we have to embrace technology if we are to face the challenges our education system will encounter in the future.

Caroline Keep – Maker Educator
@Ka81

Ahrani Logan

'EdTech is an exciting NS necessary part of education into the future. It offers great potential for simplifying tough concepts and amplifying engagement, for all levels, across multiple subject areas.'

Ahrani Logan – CEO & Cofounder of Peapodicity
@Peapodicity @AugmentifyIT

Mark Anderson

'For too long EdTech has been seen as a necessary evil rather than something which can help for good.

EdTech should be considered essential for all aspects of education, from reducing workload, supporting feedback, increasing productivity and efficiency, improving communication and collaboration.

It should be a no-brainer and hopefully (given how we were all placed in at the deep end, regardless of our position on our EdTech journey during the pandemic), EdTech is (quite rightly) here to stay. Thankfully this is a view which is now agreed by far more than we saw before.'

Mark Anderson – Global EdTech speaker & award-winning blogger
@ICT Evangelist

Linda Parsons

'The most effective way to educate a person is to immerse them into an experience. This stimulates understanding, increases their ability to communicate that experience to others and helps create a lifelong memory of that learning. This is the future role of technology in education. It is no longer a bolt-on; it can be used hand in hand with the teacher's passion to lead students into fostering their own love of learning.'

Linda Parsons – Digital Lead
@DigLin_

Dave Leonard

'The use of technology in all domains grows inexorably, so why should we in education choose not to benefit from its application?

Education needs to progress from the Victorian model that still abounds in some schools. Technology can be the key to unlock new, more contemporary pedagogies.'

Dave Leonard – Strategic IT Director
@itbadger

Paul Tullock

'I don't think there should be a separate term for technology when applied to education. I think it should be an integral part of education that underpins teaching and learning and should be integrated fully into all classrooms, schools, and all learning opportunities.

If there is one thing we have learned throughout the pandemic, it's that learning takes place anywhere, and technology makes this possible by supporting and providing opportunities, as well as providing accessibility for learners.'

Paul Tullock – Apple Professional Learning Specialist
@MrTullock

Ben Whitaker

'EdTech is never an end in itself. Technology is a tool which must always be the servant and never the master. Tech that helps teachers teach and learners learn better is the panacea – pure and simple.

Technology will never replace teachers, but teachers who use technology will probably replace those that don't.'

Ben Whitaker – Director, EduFuturists
@itsbenwhitaker

Steven Hope

'I think we need to see EdTech as an enabler for many barriers to be overcome. As an example, in further education, we have many learners who cannot access learning in a face-to-face environment at times due to childcare, caring, responsibility or other fiscal strains such as bus fare. Having cloud learning there allows learners to access learning asynchronously on any device.

In the classroom, EdTech can enhance learning when used in conjunction with high-quality learning experiences. It can be used for accessibility, to engage learners, to improve assessment and for feedback. EdTech works best when used well by the teacher and is never a magic wand – the best app we have is the teacher: they bring learning to life, and this can be with EdTech too.'

Steven Hope – Head of Independent Learning
@Hope_steven

Drew Povey

'It should be used as another tool to develop learning. I also believe that schools should prepare young people for their future careers, which will certainly involve tech, so this should be an increasing part of their educational experience.'

Drew Povey – Leadership Specialist
@DrewPovey

Caroline Wright

'EdTech became education's unofficial emergency service throughout the pandemic, supporting school-home communications and providing digital tools to aid online teaching and learning.

It will be important that schools continue this positive journey integrating EdTech tools and support services into a blended delivery of teaching and learning to support teachers, reduce workload burdens and administration and enable students to access the very best of learning resources and inspirational educational materials.'

Caroline Wright – Director General,
British Educational Suppliers Association (BESA)
@cjpwright

Simon Blower

'It is essential and should be a key part of every school's development plans!'

Simon Blower – Co-founder Pobble.com
@SimonPobble

Dan Fitzpatrick

'When we consider the future role of EdTech in education, we must firstly consider the impact of technology on how we live and work. It's huge, isn't it!

So why wouldn't we use innovative technology in education and why wouldn't we educate our learners in it? Innovation and technology go hand in hand. If we want to be innovative educators and we want our learners to be innovative, then technology must be integral to how we educate.'

Dan Fitzpatrick – Secondary School Senior Leader
@Danfitztweets

Dr Helen Price

'The provision of remote learning during the pandemic has taught us that it is possible to provide high-quality educational opportunities by effectively utilising EdTech.

However, doing so has meant many organisations have had to seriously up their game and have realised that their pre-pandemic approaches lacked strategic vision and ambition.'

Dr Helen Price – Executive Headteacher
@HATCEO

Olly Lewis

'EdTech is full of opportunity. Whether it be to streamline workflows, enable access for stakeholders, involve and support wider community initiatives, foster learning partnerships without the issue of geographical location(s), reduce teacher workload, drive personalised learning for students...the list of opportunities is almost endless.'

Olly Lewis – Educator and School Leader
@OLewis_Coaching

Abid Patel

'EdTech is one of the most vital tools for an educator to help enrich learning. Our children are now born into a digitally rich world and are surrounded by tech from the minute they are born. As educators, we are in the perfect place to help prepare them for technologically rich lives.'

Abid Patel – Trust IT Director
@Abid_Patel

To me, EdTech means...

'...the opportunity to transform educational learning experiences.' **Bukky Yusuf**

'...something that helps me to better serve my students and school community.' **Mark Anderson**

'...opening up a world of creativity and knowledge for students and freeing up school leaders and teacher time to focus on teaching (using a whole range of resources blended, print and digital) to inspire and guide pupils to achieve the best of their potential.' **Caroline Wright**

'...revolution.' **Caroline Keep**

'...using a range of technology to stimulate and enhance the learning experiences of our children as well as supporting teachers, with a focus on reducing workload.' **Becky Waters**

'...preparing kids to be future ready in a fast-paced, technological and STEAM-focused world.' **Ahrani Logan**

'...learning with the tools we will need in our future.' **Scott Hayden**

'...technology use in education, whether it be to learn about technology, to learn using technology or to learn where learning is enhanced/supported by technology.' **Gary Henderson**

'...reduced workloads, greater efficiency and maximised budgets. EdTech in the learning journey of a student can be a means of expression, a window to the world and an opportunity to connect.' **Dan Fitzpatrick**

'...a more efficient way of communicating to aid lesson planning. A transparent way for parents, students and teachers to see each student's strengths and improvements. The medium that students will use to create their own content to display their abilities.' **Linda Parsons**

'...the use of innovative technologies to enhance great teaching.' **Dave Leonard**

'...accessible learning opportunities for all, fostering independence, curiosity and creativity.' **Paul Tullock**

'...education through, with and about tech.' **Drew Povey**

'...using technology effectively to support brilliant teaching and learning.' **Simon Blower**

'...the means for helping to prepare our children for rich digital futures.' **Abid Patel**

'...start-ups or non-profits looking to scale quickly with an aim to dismantle some of the traditional educational paradigms, many of which have been in place for hundreds of years.' **Katie Stickley**

'...using technology to enhance the learning experiences of students.' **Steven Hope**

'...the tools to communicate, to collaborate, to research and to create.' **Martin Bailey**

When used appropriately, the right EdTech can...

'...transform the way that teachers teach, and the way that learners learn, by supercharging the vital human aspects of education. Technologies like AI enable educators and students to perform even better in their respective roles. Technology will never replace teachers – but it can make them far more effective, reduce their workload drastically and help them to provide their students with an even better education.'

Priya Lakhani OBE

'...transform the learning experience, accessing areas of learning and cognition that other tools cannot.'

Ed Fairfield

'...empower a child to feel incredible.'

Philippa Wraithmell

'...give students the abilities and tools to change the world we live in and without exhausting ourselves in the process.'

Caroline Keep

'...give back to teachers the cherished time needed in lessons to actually have a meaningful conversation with their students (instead of ploughing through endless and robotic mark schemes).'

Linda Parsons

'...enhance and enable learners. It can extend learning beyond the four walls of a classroom and four walls of buildings. Learning can happen any place, any time, on any device.'

Steven Hope

'...help make school systems and processes more efficient and effective, help inspire learners through access to knowledge and creativity, and support teachers and school staff in communicating effectively with parents and their wider communities to aid and support learners in their care.'

Caroline Wright

'...break down barriers and make the curriculum accessible to all.'

Martin Bailey

'...facilitate self-quizzing and distributed learning – two strategies proven to be very effective. This can be achieved through gamification, for example. Empower parents and teachers and schools by providing insights and feedback using big data.'

Katie Stickley

'...change lives. It makes learning accessible for all and provides a personalised pathway to learning. It can literally transform lives, as I've found out through my own teaching experiences.'

Paul Tullock

'...augment and enhance learning to prepare us for the expectations of future industry.'

Scott Hayden

'...increase engagement, making learning easier for students and teaching easier for teachers.'

Ahrani Logan

'...(to steal partly from the SAMR model) build, enable, enhance and allow for the reimagining of learning and the educational experience.'

Gary Henderson

'...becomes the learning tool that empowers innovators, creates life-long learners and opens up the world to digital explorers of all ages.'

Dan Fitzpatrick

'...transform lives.'

Mark Anderson

'...make a real difference to all students. As a former school leader and the parent of a child with SEN, I believe EdTech has a key role to play in ensuring the needs of all children can be met.'

Simon Blower

'...revitalise the learning experience for children, helping them to learn in newer, individual means that will inevitably raise attainment for every single child.'

Abid Patel

'...provoke, encourage and enable ALL learners to get the most out of lessons whilst saving time and effort for educators.'

Dave Leonard

'... empower learners to find a voice, encourage teachers and learners to collaborate and make everyone's jobs more efficient!'

Ben Whitaker

How should we adapt our thinking around EdTech?

Philippa Wraithmel

'We should consider that less is more. We do not need all of the shiny new tools, just those which enhance learning and support the effort to find the misconceptions in learning, the holes where students are not able to connect the information, allowing them to make progress. We should adapt by thinking that we do not always need to use technology. Just because we have it and we can use it doesn't give us our why. If we do not have a why, we reduce the impact which it can have on student learning.'

Philippa Wraithmell – Director of Digital Education & Innovation (@MrsWraithmell)

Mark Anderson

'Industry has long understood the need for agility. Education and the use of technology in schools need to accept this new requirement and become more flexible and agile.

We need to accept that technology change, despite what well-presented case studies would have us believe, seldom runs as planned and that there will always be things we haven't appreciated in advance along with missteps along the way.'

Mark Anderson – Global EdTech speaker & award-winning blogger (@ICTevangelist)

Caroline Keep

'Growth mindset. It ain't (so much) what you do, it's the way that you do it! Truth is, like many things, EdTech doesn't always work properly. It's how we respond to it and use it for our benefit (and stick with it!) and that's what gets results.'

Caroline Keep – Maker Educator (@Ka81)

Ahrani Logan

'EdTech should be thought of as a natural progression in education, recognising that students are increasingly tech savvy and that the world is not going to become less tech-focused, but more.'

Ahrani Logan – CEO & Cofounder of Peapodicity (@Peapodicity @AugmentifyIT)

Scott Hayden

'As a sector, we're too often doing half the job. Speccing and supplying, but failing to encourage user adoption and embed into the day-to-day.'

Scott Hayden – Teacher and Digital Innovation Specialist @ScottDHayden

Gary Henderson

'Stop talking about ourselves like we are separate from learning – it others the term and creates a distancing effect for those less confident to try new things.'

Gary Henderson – Director of IT @Gary Henderson18

Drew Povey

'I'm a passionate believer that education should prepare young people for the future...and the future will have tech front and focus...so what are we waiting for in education?'

Drew Povey – Leadership Specialist @DrewPovey

Simon Blower

'Ensuring teachers have the right level of support to introduce and embed new technology is key. At Pobble, we provide support to those leading Pobble and free training at any point in the year for all teachers.'

Simon Blower – Co-founder Pobble.com
@SimonPobble

Bukky Yusuf

'Work from the model of the lowest common denominator e.g. if a student or staff member does not have a device or WiFi, how can they still use the EdTech that is being provided (avoid working from the assumption that everyone has access to devices, internet etc.)?

Also, consider what can be done to support the training needs of school-based staff that are extremely busy and very short on time to learn new tools.'

Bukky Yusuf – Senior Leader and Educational consultant
@rondelle10_b

Linda Parsons

'Timely and student-focused education for social media and gaming is paramount. It is not enough for schools to "ban the phone" or blacklist social media websites/apps.

This thinking means students engage with these activities completely unsupervised, instead of teachers having the opportunity to guide students. Our online communication is just as important in everyone's lives now, regardless of age or background.'

Linda Parsons – Digital Lead
@DigLin_

Paul Tullock

'We need to remember that EdTech is education. Simple as that. The world is digital, and these transferable skills are needed across all employment from retail to medicine and beyond.

Where do we go when we want to learn how to do something or new skills? Self-paced YouTube and Google searches.'

Paul Tullock – Apple Professional Learning Specialist
@MrTullock

Ben Whitaker

'We need to operate cloud-first in our decisions as stakeholders in education and look at device and ecosystem agnosticism – people need freedom to pick and choose rather than the dichotomy of being a Microsoft or a Google school, or iPads vs Chromebooks. Collaboration is key.'

Ben Whitaker – Director, EduFuturists
@itsbenwhitaker

Steven Hope

'I think we need to use technology in different ways from just thinking of pigeonholing it into the current way we deliver some face-to-face learning. As an example, project-based learning models and using EdTech to collaborate and come together to create resources to demonstrate knowledge and skills would be great.'

Steven Hope – Head of Independent Learning
@Hope_steven

Olly Lewis

'All too often, educators have focused on the tool rather than the pedagogy and intended learning outcomes. The recent move to remote education has been a blessing in many senses as it's realigned our foci for purposeful and successful learning supported by EdTech.'

Olly Lewis – Educator and School Leader
@OLewis_Coaching

What should EdTech vendors be doing to support schools?

'Through their work with schools and education systems around the world, EdTech vendors have a wealth of knowledge of and advice for what works well in schools. EdTech suppliers are keen to work hand in hand with schools to share knowledge and support schools. EdTech vendors must continue to have access and opportunity to work with schools collaboratively to help develop solutions to new and emerging challenges facing schools in the future.'

Caroline Wright

'Training continues to be so important and should be a key part of any EdTech company's offer.' **Simon Blower**

'EdTech vendors need to be working with schools to find out what schools really need from a product. There have been so many times when I've been given the hard sell on a product that claims to be the answer to all my problems, but when I ask about the reporting side of it and if it can produce detailed reports on user statistics, I usually get told that it's in development and that will be the next thing they look at. Why not work with schools and leaders to begin with to really find out what we need, rather than assuming (from a very limited knowledge base) that what they have created is what we need.' **Jon Tait**

'Offer more collaborative "personalised" learning. More SEL and mental health support in the fabric of the curriculum and EdTech. More integrated SEND tools and training. Push the case for continuous assessment. Build sector voice for going beyond "content dumping" into best practice with use of EdTech showcases: how EdTech can connect learners with experts, P2P learning in their areas of interest.' **Sophie Bailey**

'Listening to the needs of each region and how this would impact end usership, as the social climate in the UK is different to Finland, which is different to Bahrain, etc.' **Linda Parsons**

'EdTech vendors could support schools more by offering high-quality, cost-effective solutions that allow all schools to reach a similar and higher level of EdTech teaching. Being aware of pricing should be a key part of enabling equity of the learning experience for students across schools.'

Ahrani Logan

'Create tools which are needed, not use tools and turn them into education solutions.'

Philippa Wraithmell

'It should be a partnership, not just about making sales.'

Mark Anderson

'Building great working relationships and providing continuing support to customers rather than just taking the money and running/box shifting. Too many vendors play the percentage game of going in high on price in the hope that educators will be too busy to question the price they are paying. Be transparent, be honest and be there when we need you. We get that you have to make a profit, but we need to work in partnership to grow together.' **Dave Leonard**

'EdTech vendors should ensure there are adequate trial periods for schools to assess whether their product is fit for purpose before purchasing. When purchasing larger hardware (e.g. touchscreens), there should be: a bespoke staff training programme included within the cost with the choice to update training on a regular basis; follow-up consultations with key members of staff to ensure the product is being used to best effect; updates (e.g. termly email) which can be shared with staff to remind them of the benefits of using the technology and give tips/hints.' **Becky Waters**

'Be clearer on how they manage and use the data they gather, but in clear, concise and easy to understand language. In addition, make it easier for schools to discharge their GDPR duties in relation to using third parties by providing clear answers to the basic privacy impact assessment questions which schools should be conducting. Provide more realistic examples of the use of their products, including where errors or missteps have been made, instead of case studies which underrepresent these.'

Gary Henderson

'Work with teaching and learning teams and NOT IT systems teams.' **Scott Hayden**

'More! Schools need to insist on and expect higher standards of service and support. From up front accurate advice to ongoing training and CPD.' **Ed Fairfield**

'Continue the conversations and build a relationship with schools. The best now continue to chat away and engage via social media.'

Martin Bailey

'Support the design and implementation of short-, middle- and long-term plans for EdTech solutions as this will help build a more strategic overview. They should also help schools in reviewing the use of EdTech solutions from different perspectives (e.g. school leaders, class teachers, students, parents) so that impact can be measured when considered from these perspectives.' **Bukky Yusuf**

'Making it easy to integrate with whatever device or ecosystem has been chosen.' **Ben Whitaker**

How should the government be supporting the development of technology to support teaching and learning?

'Rethink assessment using EdTech.'

Sophie Bailey

'It will be important that future government interventions recognise and celebrate the expertise and autonomy of school leaders and multi-academy trusts in supporting their school communities, alongside the innovation and creativity of the UK's vibrant EdTech sector. Lockdown has proved that there is no one-size-fits-all for education. Government should support and champion school choice rather than imposing top-down centralised solutions and schemes.'

Caroline Wright

'Government should be taking a more proactive interest in speaking with smaller, more innovative EdTech providers to find cost-effective solutions to support technology development in schools.'

Ahrani Logan

'Shine a light on bright spots at lesser-known schools and colleges. The usual examples we see are impressive, but they are so good that it is intimidating to many. We need to see smaller, more achievable examples, so they become more tangible.'

Scott Hayden

'Funding is critical for schools to make use of technology. Only with the fundamental building blocks such as internet access, infrastructure, devices and support can schools then seek to make use of technology in lessons and for learning. I also believe the government could have key roles in supporting schools developing their strategy and in supporting teachers and students in using technology.'

Gary Henderson

'Making sure that they fix their own house. I'm talking about Ofsted. Making sure that schools are given the freedom to adapt learning and embrace the use of EdTech in teaching and learning, as opposed to penalising schools for being brave to take risks.'

Abid Patel

'Nesta does a good job, but its projects are a little narrow. To be involved, it's a matter of who you know. The provision of devices and internet access is the biggest way that they can support learners, whilst repairing the damage that significant cuts to school budgets have done in the past couple of decades would be of great benefit to the in-school experience of both teachers and learners.'

Dave Leonard

'Firstly, it should be tied directly to ITT. Secondly, it should learn from the benefits seen by Wales and Scotland from having a unified approach. And thirdly, it should have proper funding.'

Mark Anderson

'1:1 for all students. Equitable access to data and high speed WiFi. opportunity to develop blended learning models as SOP rather than just as a response to COVID-19. Self-paced learning models for every age range – similar to Open University. More inclusive demonstrator-style programmes.'

Ben Whitaker

'An office for EdTech with clear goals for education. We have an office for artificial intelligence, yet not one for education. How's that work?'

Caroline Keep

'More support for small start-ups. Some of my favourite EdTech products are not from the huge companies.'

Martin Bailey

'Everyone should have access to a personalised device and schools, staff should be fully supported and trained and infrastructure should be up to standard.'

Paul Tullock

'Allowing infrastructures to be developed as part of education strategy, to be giving the support and tools to ensure that every school has the tools at their fingertips and not have to spend budget after budget to try to be part of the global EdTech game. Also, we should be moving towards embedding digital literacy into ALL subjects and safety into the core pastoral education system. We have to stop pretending it is one person's job – generally the IT department's. It is a bigger concept; it is everyone's responsibility.'

Philippa Wraithmell

'I think investment and look back at agendas such as FELTAG and ETAG and look at how those can be used as a framework for moving things forward.'

Steven Hope

'I believe they are starting to do some of the following; BESA LearnEd roadshows Creation of TECH UK and Digital Capital Testbed schools to test and pilot technology helping both schools and companies. So in short; more partnership programmes like those above.'

Katie Stickley

'Whilst the provision of IT equipment was available during lockdown, not all schools were allocated devices (e.g. infant schools were overlooked!). All children need access to a good quality device and a reliable WiFi connection in order to make using technology with children at home as effective as possible. To achieve this, dedicated funding is needed. Furthermore, all schools need dedicated funding for the development of technology within school. Purchasing good quality hardware and software should be a priority, but sadly this is not possible for some schools due to diminishing budgets.'

Becky Waters

Thoughts on the need for a digital strategy

'I think the community needs to be part of the digital strategy. I think education needs to be passed through the school to parents and carers at home. The common issues should be discussed across the region/area/local schools; it is likely that if something digitally is happening in one school, it will also be happening in another. Sharing this information is key to educating everyone to be safer and stronger. We need to remember that many adults are digitally illiterate, even though they have devices etc. They will do their best to support their children, but if they do not understand it themselves, how can they keep up to date without the support of the schools and communities.'

Philippa Wraithmell

'We have a digital pathway and strategy that has been critical to our school being able to open our makerspace and run digital enrichment. Having a clear vision is so important. You wouldn't plan a lesson without having a scheme of work, a curriculum roadmap, etc. Digital is no different.' **Caroline Keep**

'Just like a teaching and learning policy or a parent-teacher agreement, a digital strategy welcomes everyone into the school community and makes everyone feel like they have a valid voice. Without a digital strategy, schools will find it incredibly hard to convince students, parents and teachers to change ingrained habits. This may lead to the school overspending in time and money to integrate a product. and having to compromise on their long-term vision.' **Linda Parsons**

'I completely agree with the need for a digital strategy. A wise man (Al Kingsley) once likened this to a car journey. If you don't know where you are and where you want to get to, how can you know that you are going in the right direction?' **Dave Leonard**

'It is important to have a digital strategy. With one in place, a school and its students, staff and parents know where the school is going with its technology use and why specific tools are being used. As Covey would say, "Begin with the end in mind." There needs to be a shared understanding, and having a digital strategy is a key part of developing such a shared understanding.'

Gary Henderson

'I would agree that strategies at a macro level would foster more collaboration across schools and would rightly force the issue of digitisation onto county, academy or school development plans. This would help to give EdTech the impetus it urgently requires and deserves.'

Katie Stickley

'I think the culture and readiness to innovate has to be there, and a strategy to guide and support this around developments would be great.'

Steven Hope

'At the Digital Citizenship Institute, we believe creating a digital strategy needs to be an experience, not just an event or an assembly. The ideal strategy should also involve the entire school community. Professional development is an extension of being a lifelong learner and needs to include learning for all ages at school, home, play, and work. This intergenerational approach reminds us that it doesn't matter how young or old we are, using technology to solve real problems and create solutions is everyone's responsibility.'

Marialice B.F.X. Curran, PhD

'My experience is that many schools create their own strategy, and in many cases (from my own experience) one member of staff drives this. It should be a team effort led from the top by strong leaders who know the possibilities and invest for the future. Schools and districts need to know what is possible from the leaders and the different directions it can take and how these journeys look.'

Paul Tullock

'We are fortunate that AI has helped us to develop our digital strategy, which is still a work in progress. A key strand in this work has been planning the professional development strategy for school staff, to ensure that they are increasingly confident to use EdTech to enhance learning.' **Dr Helen Price**

'Agree. Many schools don't plan for "a cycle" and then it becomes a huge knee-jerk outlay, rather than a constant refreshing of technology.' **Martin Bailey**

'The digital strategy and the teaching and learning strategy need to be created alongside one another if they are not the same thing.' **Scott Hayden**

'I've always said that schools have strategies for undertaking every challenge they face. Whether it be teaching and learning, safeguarding, finance, Ofsted. There are strategies in place to help tackle each of these key areas of focus. EdTech is not a utility like electricity, gas and water. Each of those key areas probably make heavy use of EdTech and IT. Without the IT, they will all probably fail. This is why schools need a digital strategy: to ensure that EdTech is looked at right at the core and that all key functions of school life are built around the understanding of how much of an important role EdTech can and does play.' **Abid Patel**

Can EdTech support staff and student wellbeing?

'It's crucial – digital wellbeing is the focus of my career now.'

Scott Hayden

'Wellbeing is integral to learner success, so it needs to be weaved into EdTech development and the school curriculum.'

Sophie Bailey

'Excited by the potential of voice-to-text and verbal and video feedback. Great time saving benefits. Making yourself a priority is not selfish; it is the best way for others to benefit from your skills too.'

Martin Bailey

'Without "well" we can't "be". Wellbeing underpins all aspects of life; only if it is right can you succeed, flourish and thrive. EdTech has potential. Personalised feedback and interaction, reduce staff workload and admin time.'

Paul Tullock

'Rule 1 of teaching is about ensuring the safety of the students in your care. If their wellbeing is low because of something your school is doing, then you are doing the wrong things and you must re-evaluate. I think that we can do a lot to support the workflows within our schools to ensure that there is no unnecessary work going on in schools. But we must be mindful that the additional time gained is not there for staff to be able to then do more work; it is there so they can be more balanced. We also need to be able to use it to be more aware of what we do and how much we use. I think things like building in screen time checks can be very useful – asking ourselves to make sure it is only used when needed, sometimes even having no-device days.'

Philippa Wraithmell

'It can play an important part yet is underutilised in this regard. There are more apps that now address this; however, without being linked to whole-school improvement plans or a whole-school focus, it runs the risk of having minimal impact. Also, there needs to be a recognition that different people (within the same school/school community) will require a different focus for their wellbeing needs.'

Bukky Yusuf

'Teaching and learning is a social experience, and our teaching and learning is best when we are content/happy. As such, wellbeing is critical to the central aim of education, teaching and learning in schools. There is a key need to examine and discuss the balanced use of technology – where it can be beneficial but also where it represents a risk or may cause harm.'

Gary Henderson

'EdTech can greatly help educators to streamline, simplify or even remove altogether some of the most mundane tasks. Feedback and assessment has been revolutionised from boring lines of red text into rich snippets of valuable information that can really help to reshape the lives of the children receiving such personal and individualised support from their teachers.'

Abid Patel

'I think things can be made more efficient with EdTech and tasks can be made smoother and easier. I think we also can push the importance of purposeful screen time and device use. Wellbeing is key and fundamental to healthy learning environments. We have to be shown that we need to look after ourselves first and then take care of others too.'

Steven Hope

'Wellbeing must always come first. If the person behind the technology isn't healthy, then neither is the technology.'

Dan Fitzpatrick

'Digital wellbeing is at the foundation of a school community. What are some ways we are actively supporting wellbeing both online and offline for our entire school community? During remote learning, there has been a lot of conversation around etiquette regarding cameras being on and students being present online, but not enough conversation has happened about creating safe online spaces where both teachers and students have a system in place to let others know we need time away from a screen. The need for balance and time to completely unplug must play a role in how we learn, play, and work now.'

Marialice B.F.X. Curran, PhD

'Anything that reduces the time it takes to perform onerous tasks is a huge help to wellbeing. EdTech can free up the most precious commodity – time.'

Dave Leonard

'I think the best way to develop "well beings" in our school is to arm them with tools to help address their individual and specific needs. I believe from personal experience that tech is by far the best, quickest and least intrusive way to achieve this! For too long, people have talked about health and safety...with the sole emphasis on the safety. Let's address the health agenda and bring more happiness to our lives, both personally and professionally.'

Drew Povey

'Screen time can be problematic for this but digital tools like Headspace and step counters are a really good opportunity to help wellbeing. Screen fatigue is real, and we must do more to protect work-life balance. When managed effectively in terms of notifications and cultural expectations of working hours, EdTech can be a huge time saving tool, which is what teachers are short of.'

Ben Whitaker

'Technology is ingrained in every part of our lives, and access to information is now at everyone's fingertips, regardless of the quality and quantity of this access. Schools must have strict and clear rules about their contactable hours to protect parents, students and staff wellbeing. Used correctly, the insight options on some platforms can be used to monitor and start a conversation about a student's attendance and hours online within education platforms. This will not only help the students that are getting lost and are not attending class but also help the students that are spending hours and hours on their work, at the expense of getting quality rest time.'

Linda Parsons

On EdTech supporting greater parental engagement

'EdTech has facilitated better use of parents' evenings, more data to give clear pictures of students' progress and how we can embed interventions. Webinars to share messages – these can be re-watched after for those parents who cannot get to them on time. Teaching them the skills as well: bridge the gaps!'

Philippa Wraithmell

'Digital parents' evenings have been a huge success from the perspective of both teachers and parents. Whilst face-to-face meetings are still valuable, the use of EdTech will allow communication to take place on a more regular basis and enable those working during the hours of parents' evenings or with other barriers to their attendance to speak to their child's teachers and support their education.'

Dave Leonard

'It allows interaction with teachers, a way to share stories from home and school that benefit the pupil. It allows the chance to support our children better and build learning bonds as a family. A glimpse into the life of a child (not just the learning) shows how we can support each and every pupil better.'

Paul Tullock

'Lockdown has brought teachers and parents closer together. This is good, although there are risks around direct uncapped access to teachers by parents. Managed tech comms platforms can help.'

Ed Fairfield

'One of the key benefits with the use of technology is how it has enabled communication, from email, through social media and Facebook, to the recent growth of Zoom during the pandemic. Schools can therefore look at how the various technology communication channels can be used to engage parents.'

Gary Henderson

'I believe that the "education triangle" of parents, students and staff have all benefited hugely from seeing how we can operate a learning environment in a remote world. The way everyone adapted was truly incredible!' **Drew Povey**

'As part of our EdTech innovation grant, we have looked carefully at everything from ensuring parents have the right access to our platform through to supporting them with celebrating their children's writing achievements.'

Simon Blower

'As a parent myself, I have always struggled to understand how and what my children were learning. The remote learning experience has really helped me to understand what my children are learning and enabled me to support my children with their ongoing learning. This was really difficult pre-COVID without the EdTech. Now, my children not only have a clear view of their learning but also have a direct line to their teachers outside of the classroom, if they need further support.' **Abid Patel**

'EdTech has enabled remote parents' evenings – this is a fantastic example of EdTech supporting wider aims of student development while involving all stakeholders.'

Olly Lewis

'Many EdTech platforms can be used by parents to monitor their children's progress over time and support their needs both academically and emotionally. Data is easily available and accessible to parents in a way that it never has been before. However, overanalysis of a child's performance could also lead to problems including low self-esteem and poor mental health – so everything in moderation.'

Katie Stickley

'Who would have thought something as simple as online parents' evenings would get people excited about some legacies of online learning? Building on this, we need to see past the building. Parents also inhabit the virtual world, so let's meet them there.' **Dan Fitzpatrick**

The importance of digital citizenship in schools

Marialice B.F.X. Curran

'We view digital citizenship as an action, something that we need to practise and do every single day.

In today's interconnected world, this is our opportunity to put global education into practice to empower others to become change makers for using tech for good in local, global and digital communities.'

Marialice B.F.X. Curran, PhD – Founder & Executive Director, Digital Citizenship Institute @DigCitInstitute

Philippa Wraithmell

'Digital citizenship is so important – we live in a world which is far too digital to not be addressing this in all schools, hoping that people will just learn. I think the fear of delivering it properly comes from teachers' lack of knowledge. Train them ALL in digital citizenship and allow them to be empowered. This will also help them to be able to spot students who may be at risk of being online and falling into traps.'

Philippa Wraithmell – Director of Digital Education & Innovation @MrsWraithmell

Gary Henderson

'Digital citizenship is critical in schools to ensure that staff and students both understand the benefits and potential for technology but also the risks and the dangers. This includes the basic eSafety discussions, but increasingly discussions on big data, individual privacy vs public good, bias and influence, etc.'

Gary Henderson – Director of IT @Gary Henderson18

Dave Leonard

'Pressure on the curriculum has led some schools to reduce the amount of discrete IT taught, and alongside this the amount of digital citizenship. This had led to a decrease in confidence and safety of students whilst online. It is key to teach digital literacy, digital citizenship and digital wisdom to young people in order to enable them to use the tools available to them effectively and safely.'

Dave Leonard – Strategic IT Director
@itbadger

Paul Tullock

'The same as citizenship. We should be behaving as we would in the real world. We need to be accountable for our actions in person and digitally.'

Paul Tullock – Apple Professional Learning Specialist
@MrTullock

Steven Hope

'Learners need to know how to have positive regard for others and best practice, whether this be in person or in the digital sphere.'

Steven Hope – Head of Independent Learning
@Hope_steven

Dan Fitzpatrick

'Digital citizenship is vital. At a young age, we are shown how to interact with others and the dynamics of a classroom. If we want our learners to be people who develop independent learning skills, then they need to be shown how to interact online and the dynamics of the virtual world. We are failing them if we are not doing this now.'

Dan Fitzpatrick – Secondary School Senior Leader
@Danfitztweets

Abid Patel

'Tech is all around us. Like most things in life, it can be used for good and for bad. As educators, we play a very big role in ensuring that we help our children to understand how to be model digital citizens and use tech to do good, innovative things that will help to improve the lives of everyone around us.'

Abid Patel – Trust IT Director
@Abid_Patel

Caroline Keep

'If we don't prepare our pupils with digital citizenship, we fail them in preparing to navigate life outside our school gates.'

Caroline Keep – Maker Educator
(@Ka81)

Key considerations when managing your IT infrastructure

'There is a constant battle between security, usability and cost. It is imperative to manage these three elements whilst delivering a safe environment for teaching and learning and making progress towards the goals identified in your organisation's digital strategy.'

Dave Leonard

'We need infrastructure which is flexible to support changing demands, sustainable (both financially and in terms of support and the environment), and secure.'

Gary Henderson

'Cloud-first. Device-agnostic. Equitable access to tech and internet.'

Ben Whitaker

'Like satnav, assess where you are. Plan where you want to go. Then plan the route. Don't just switch on and start driving.'

Ed Fairfield

'Ensure it is future-proofed and pushed for big capacity, not just right for now but for future developments, so it is robust and ready now.'

Steven Hope

'Having your IT team on your SLT team, making sure they understand education as well as technology, having key links with these teams – these measures will help them to understand how and why things need to be in place. They should not just be there to check interactive boards work and computers. We must put them on safeguarding courses and integrate them into the school properly. This way they can add in the key technological components into your schools, have a better understanding when people request platforms and be able to develop safer work systems.'

Philippa Wraithmell

'Get the planning right from the start. You wouldn't buy a car without an engine then add it as you went, so you shouldn't buy devices without the infrastructure to manage them and have them operate at full capability (WiFi, network, storage etc.). Teachers should be external and focus on teaching and learning, letting the technicians know what they want and need so they can work together as a team. Too often the technician drives the strategy or need and not the teachers. Everything needs to be in place to succeed.'

Paul Tullock

'Listen to your IT directors/managers/technicians. They are the people on the front line, doing a very tough job, usually in very lonely, unappreciated roles, with very little support and understanding of the challenges they face every single day. Connect with them, make them feel appreciated and always ensure to include them in planning discussions from the beginning, not introduce them like an afterthought. Treating your IT teams with respect will ensure that you reap the rewards of making EdTech work for you.'

Abid Patel

Other Quotes

'Throughout the history of education, there have been many ways to teach one thing, and you should pick the way that suits you. So, if your neighbour is using something different to you, take the time to listen to them. You might find your new favourite tool!' **Linda Parsons**

'We need to be methodical, iterative and collaborative to make this work globally. To support all educators. Do not rely on COVID-19 to be the push; we need to keep the momentum going and keep it relevant and engaging.'

Philippa Wraithmell

'The pandemic has brought about the largest period of change in education in my 20+ years working with schools/colleges. We now need to reflect on the changes over the last year and take from it as many positives as possible and resist the urge to return to how things were, with nothing gained. We also need to accept that the future is likely to see increasing uncertainty, so we need to be prepared to continue to be flexible and ready for change. I hope that the last year has highlighted the important part of technology in education and therefore will lead to a new focus and investment in technology for use in teaching and learning.' **Gary Henderson**

'We need to catch up to other industries, then blaze the way. These industries need learners and skills, and until we invest properly, we will always be behind and plugging gaps.' **Paul Tullock**

'At a time when physical visits and visitors have not been possible, I have loved how education technology has still been able to bring the curriculum to life for pupils.' **Martin Bailey**

'The EdTech landscape holds great potential for students, teachers and parents. Making the next generation future-ready requires that the current generation of knowledge-givers keep up with the evolving STEAM landscape. We created AugmentifyIT as a knowledge-rich, easy to use, purse-friendly AR, to bring memorable learning experiences and generate a love of STEAM in students, and for them to take that into their futures.'

Ahrani Logan

'Schools need vastly better support to overcome an array of EdTech challenges. In amongst all the other carnage schools need to contend with. The good news is COVID-19 has dragged EdTech from the neglected 'important/not urgent' category to front and centre. Now is the time to embrace it in full.' **Ed Fairfield**

'We have to develop learners' skills to be ready for learning now and then progress too. Skills such as creativity, critical thinking, communication and collaboration are key and should not be an add-on or 'either/or' with subject knowledge – both are vital.' **Steven Hope**

With my huge thanks to the following people for their insights and support

Abid Patel, Trust IT Director, @Abid_Patel
Ahrani Logan, CEO and Co-Founder, Peapodicity, @Peapodicity
Becky Waters, Headteacher, @SpoonerWaters
Ben Whitaker, Director, EduFuturists, @Itsbenwhitaker
Bob Harrison, Visiting Professor and Educationalist, @BobHarrisonEdu
Bukky Yusuf, Senior Leader and Educational Consultant, @Rondelle10_b
Carl Ward, Chair, Foundation for Education Development, @FedEducation
Caroline Keep, Maker Educator, @Ka81
Caroline Wright, Director General, BESA, @CJPwright
Dan Fitzpatrick, Secondary School Senior Leader, @DanFitzTweets
Dave Leonard, Strategic IT Director and www.Learningdust.com, @ITbadger
Dr Helen Price, Trust Executive Headteacher, @HATceo
Drew Povey, Leadership Specialist, @DrewPovey
Ed Fairfield, EdTech Advocate, @Mreddtech
Gary Henderson, School Director of IT, @Garyhenderson18
Jon Tait, Director of School Improvement and Deputy CEO, @TeamTait
Katie Stickley, Co-Founder, DidTeach and T.I.M.E., @DidTeach1
Linda Parsons, School Digital Lead, @DigLin_
Marialice B.F.X. Curran, PhD, Founder and Executive Director, Digital Citizenship Institute, @Mbfxc
Mark Anderson, ICT Evangelist, Education Lead, NetSupport, @ICTEvangelist
Martin Bailey, Digital Enrichment Leader, @LanchesterEP
Olly Lewis, Educator and School Leader, @OLewis_coaching
Paul Tullock, Apple Professional Learning Specialist, @MrTullock
Philippa Wraithmell, Director of Digital Education and Innovation, @MrsWraithmell
Priya Lakhani, CEO and Founder of Century Tech, @Priyalakhani
Scott Hayden, Teacher and Digital Innovation Specialist, @scotthayden
Simon Blower, Co-Founder, Pobble, @SimonPobble
Sophie Bailey, Founder and Host, The EdTech Podcast, @Podcastedtech
Steven Hope, Head of Independent Learning, @Hope_steven

10

Places to learn from – my top 10s

Who doesn't love a top 10? Like you, I could probably create 50 top 10s, so I am mindful that there are many more options than those I have included here. Please (please) don't be offended if you're not on the lists – that is absolutely not my intention.

Blogs

Blogging and educators are like peas in the proverbial pod nowadays and it's the continual willingness to share with peers that makes the profession stand out from most others. I do partake in blogging myself on a regular basis, but there are some far more impressive blogs out there that are a fantastic source of ideas and advice from people I trust. I couldn't possibly list them all – it is, after all, just a top 10 – but these are worth checking out.

Class Tech Tips,
Monica Burns
www.classtechtips.com

Scan me

Shake Up Learning
Kasey Bell
www.shakeuplearning.com

The ICT Evangelist
Mark Anderson
www.ictevangelist.com

Teacher Tech
Alice Keeler
www.alicekeeler.com

CoolCatTeacher
Vicki Davis
www.coolcatteacher.com

eSafety Advisor
Alan Mackenzie
www.esafety-adviser.com

Deigned to Teach Digitally
Philippa Wraithmell
www.designedtoteachdigitally.com

DigiLin Learning
Linda Parsons
www.digilinlearning.com

The Edvocate
Matthew Lynch
www.theedadvocate.org

The Innovative Educator
Lisa Nielsen
theinnovativeeducator.blogspot.com

Podcasts

Those who know me will also know I do enjoy taking part in a good podcast. But as much as I love to participate myself, I'm also a regular listener of many amazing shows that feature a wealth of inspiring guests and, of course, some great hosts. More than ever, we can access our learning through so many mediums and this one, for me, is one of the most accessible. I have shared below, in no particular order, my top 10.

Learning Dust – www.LearningDust.com
Summary – Education and technology go hand in hand these days, but all too often teachers and technicians don't. Here at Learning Dust we believe that wonderful learning and teaching can be achieved by encouraging techies and teachers to work together. Our motto is 'Pedagogy and Technology in Harmony', for that is how we can achieve the best results for students and staff working in education.

Our podcast aims to bring people from both camps together to discuss pertinent issues and the latest EdTech products and services. Co-hosted by Dave, an ICT Manager, and Andy, a Lead Practitioner for Teaching & Learning, we aim to make each episode entertaining, informative and thought-provoking.

Our name was inspired by a quote from Professor Tom Crick who said that 'no one ever picked up an iPad and magic learning dust fell out of it'. Our hope is that listening to the podcast could provide the learning dust that your school needs.

Edufuturists – www.edufuturists.com

Summary – From Ben Whitaker, Dan Fitzpatrick and Steven Hope. The Edufuturists podcast is a continuing conversation with guests from all over the world about education, technology, the future and everything between.

The School Leadership podcast – www.naht.org.uk

Summary – Director of Policy at NAHT and NAHT Edge Director James Bowen interviews the leading voices in education covering the topics at the heart of leadership and learning.

The Ignite EdTech podcast – @EdTech_Podcast
Summary – The Ignite EdTech Podcast is a weekly discussion around the world of educational technology, led by leading, global EdTech consultant Craig Kemp. Each week Craig gives insights into EdTech tools for your K–12 school.

He gives advice and tips on how to use technology authentically and purposefully and chats to leading educational experts and thought leaders from all over the globe.

Tiny Voice Talks – https://tinyvoicetalks.buzzsprout.com
Summary – Tiny Voice Talks gives a voice to everyone in education. In it, they speak to other passionate educators about current educational issues, key research and of course why it is so important that we continue to learn from one another.

The EdTech podcast – www.theedtechpodcast.com
Summary – The EdTech Podcast gets behind the personalities in global education innovation & EdTech. Join 1500+ listeners each week, from 141 countries, to hear from educators across early years, schools, higher & further education, plus investors, start-ups, blue chips, government & students. Also, regular insights from interviews, events and LIVE podcasts.

EdTech Talks – www.anchor.fm/edtech-talks
Summary – Phil and Chris are educators, passionate about the impact that digital solutions can have on teaching, learning and working across the education sector. You can follow the podcast on Twitter @EdTech_Talks

House of #EdTech – www.chrisnesi.com
Summary – The podcast helping educators integrate technology by sharing stories from teachers, explaining the lessons learned, and sharing actionable easy #edtech tips and tools – because whether you use it or not, technology is changing the way we teach and how our students learn.

Easy EdTech Podcast with Monica Burns – www.classtechtips.com
Summary – Easy EdTech brings you teaching strategies, tips and activity ideas. Each podcast focuses on how to make EdTech integration easy with a shout out to favourite EdTech tools for classrooms. It provides actionable, relevant tips for teachers so you can make this school year the best one yet! Join author, speaker, and blogger Dr Monica Burns for an EdTech podcast that helps take the stress out of integrating education technology into your classroom.

NetSupport Radio – www.netsupportsoftware.com/radio
Summary – Stay up to date with the latest #EdTech broadcasts including the 'Check It Out' and 'Tip Top Tips' Edu shows. Listen back to our previous podcasts on topics such as safeguarding, digital citizenship and teaching with EdTech – plus, catch up on all the activity and interviews with key education experts from across the sector.

You can also listen to chats about all things education-related from specialists in the sector in our Big Education Debates, as well as learn about how EdTech can help your school or Trust with managing its IT.

Expos and conferences

We all love a good EdTech event (don't we?) and of course some are better than others, some are more focused on solutions and others on CPD. In reality, the highlight of most events is the opportunity to catch up with peers, meet face to face with many of our online PLN and find out what's new in the sector. I've attended many events around the world and thought I'd share some of my favourites. Hopefully you recognise a few yourself.

CoSN Conference
Various cities in the US, March – www.cosn.org

The CoSN conference is organized by the Consortium for School Networking. It's a hands-on event designed to explore the latest and greatest in technology while giving school district staff the ability to meet thought leaders, try new EdTech and discuss important ideas within dozens of breakout sessions and workshops.

The GESS conference & Expo
Dubai UAE, March – www.gessdubai.com

As the organisers say:

> GESS Dubai is an education conference and exhibition with a difference – all their content is free of charge and they bring together engaging industry leaders from around the world and influential local practitioners to present a combination of inspirational talks and dynamic, hands-on workshops. The exhibition offers educationalists exposure to the latest innovations in educational products and services, as well as live demonstrations from global market leaders.

The Bett show
London UK, January – www.bettshow.com

The Bett show is hosted at the London ExCeL every January. The organisers summarise the event as being 'where technologies, practices, ideas and people come together'. Bett celebrates education and inspires future discussions as together we discover how technology and innovation enable educators and learners to thrive. It's probably the largest event of its type in the world.

The ISTE Education conference
Various cities in the USA, June – https://conference.iste.org

The ISTE conference is held annually and moves each year to a different city in the US. The International Society for Technology in Education (ISTE) is a non-profit organisation that works with the global education community to accelerate the use of technology to solve tough problems and inspire innovation. Their worldwide network believes in the potential technology holds to transform teaching and learning.

The EDUCAUSE annual conference
Various cities in the US, October – https://events.educause.edu

The EDUCAUSE annual conference showcases the best thinking in higher education IT. The organisers explain: with the best presenters, the best content, and the best networking, it's the premier higher ed IT event that brings together professionals and technology providers from around the world to share ideas, grow professionally, and discover solutions to today's challenges.

The TCEA convention & Expo

Texas USA, February – www.tcea.org

The Texas Computer Education Association convention is held annually in February. As the organisers explain:

> Since 1980, TCEA has worked to connect educators with the latest pedagogical ideas, teaching techniques, and tech tools – and with each other. Our annual convention & exposition showcases the most influential thinkers and leaders in education, along with the biggest names in the EdTech business, all sharing a wealth of knowledge with our attendees.

21CLHK Conference

Hong Kong, November – www.21clhk.org

The 21st Century Learning Conference in Hong Kong is a conference that focuses on contemporary practice and innovation in K–12 education. Whatever your role in education is, the 21st Century Learning Conference has something for you. 21CLHK draws world-renowned keynote speakers, leading educators from around the Asia-Pacific region, and K–12 educators and administrators from schools from over 33 countries.

EduTech Conferences

Global, through the year – www.terrapinn.com

Terrapinn organise the EduTech conferences around the world each year, including EduTech Europe, India, Indonesia, Australia, Africa and Asia events. Both their physical and virtual events are designed to bring educators together and introduce them to the very best EdTech solutions the world has to offer.

The CUE Conferences

California and Florida US, Spring & Fall – www.cue.org

The CUE, Computer-Using Educators, is a non-profit educational corporation founded in 1978. CUE's goal is to inspire innovative learners in all disciplines from preschool through college. With an active current membership of thousands of educational professionals, CUE supports many regional affiliates and learning networks. CUE conferences are California's premier educational technology events. CUE is the largest organisation of its type on the west coast and one of the largest in the United States.

The FETC show

Florida USA, January – www.fetc.org

The Future of Education Technology Conference is held in Florida each year. As the organisers say:

> It's the place for entire education teams to come together and learn the latest on EdTech, build a community of dedicated peers, and gain ideas to achieve classroom, school and districtwide technology goals.

Books

Books and education are almost synonymous and we are blessed to have so many innovative teachers, trainers and thought leaders who are all too happy to share their ideas and best practice. I was completely spoiled for choice in this section so I have included a few I liked that also had relevance to the purpose of this book.

How to Sketchnote: A step-By-Step Manual for Teachers, Sylvia Duckworth.
Summary – Discover the benefits of doodling!

Sylvia Duckworth makes ideas memorable and shareable with her simple yet powerful drawings. In *How to Sketchnote*, she explains how you can use sketchnoting in the classroom and that you don't have to be an artist to discover the benefits of doodling!

How to Have a Great Life: 35 Surprisingly simple ways to success, fulfilment & happiness, Paul McGee
Summary – *How to Have a Great Life* starts with you – your strengths and potential and how to develop those. It helps you understand how to tap into your ability to grow, while equipping you with insights, inspiration, and practical tools to deal with whatever life throws your way in order to achieve success and live a happy and fulfilled life.

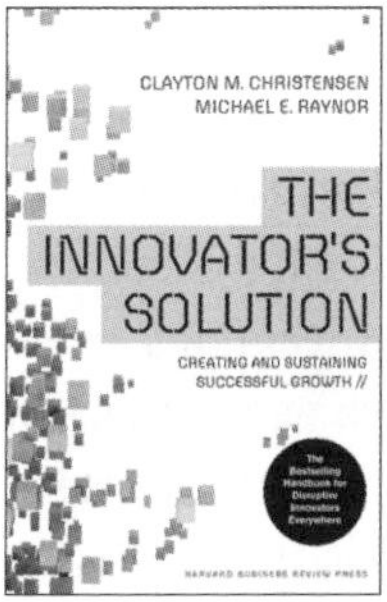

The Innovator's Solution: Creating and sustaining successful growth, Clayton Christensen and Michael Raynor
Summary – *The Innovator's Solution* expands on the idea of disruption, explaining how companies can and should become disruptors themselves. This classic work shows just how timely and relevant these ideas continue to be in today's hyper-accelerated business environment. They give advice on the business decisions crucial to achieving truly disruptive growth and propose guidelines for developing your own disruptive growth engine. The authors identify the forces that cause managers to make bad decisions as they package and shape new ideas and offer new frameworks to help create the right conditions, at the right time, for a disruption to succeed.

Rewiring Education: How Technology can unlock every student's potential, John D. Couch
Summary – *Rewiring Education* presents a bold vision for the future of education, looking at promising emerging technologies and how we – as parents, teachers, and voters – can ensure children are provided with opportunities and access to the relevant, creative, collaborative, and challenging learning environments they need to succeed.

Perfect ICT Every Lesson, Mark Anderson
Summary – [This book] uses the technology-related elements of the recent subject reports from Ofsted to provide clear and practical strategies that are proven to be successful in classrooms, offering up ideas for how they can be turned into a daily reality for all teachers.

Teaching Rebooted: Using the science of learning to transform classroom practice, Jon Tait
Summary – *Teaching Rebooted* uncovers the most important pieces of educational research on the science of learning, helping teachers to understand how we learn and retain information. Jon Tait explores strategies such as metacognition, interleaving, dual coding and retrieval practice, examining the evidence behind each approach and providing practical ideas to embed them in classroom practice.

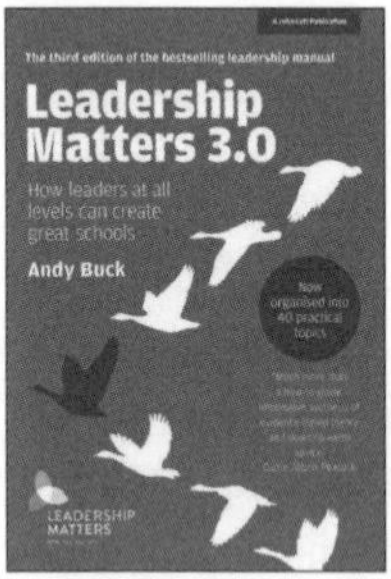

Leadership Matters 3.0: How leaders at all levels can create great schools, Andy Buck
Summary – *Leadership Matters 3.0* improves the educational outcomes for children by empowering educational leaders in national, regional and local contexts to examine, refine and develop their management skills. Andy takes an in-depth and diagnostic approach, encouraging leaders at all levels in schools to think about their own personal qualities; their specific situation; their own leadership actions; and their own overall leadership approach.

The Blended Workbook: Learning to Design the Schools of our Future, Michael Horn and Heather Staker.
Summary – Successfully implement a blended learning program with this step-by-step guide. The book supports through real-world implementation exercises how to get the most out of the ideas shared. A companion book to *Blended*.

Putting Staff First: A blueprint for revitalising our schools, John Tomsett and Jonny Uttley

Summary – Whilst it is easy to say that schools would not exist if it were not for the students, the glib converse is that without truly great school staff, the students would not be taught. What we need – as recruiting subject specialist teachers, school leaders and specialist support staff becomes increasingly difficult – is a revolution in how we treat our school staff. We have to put our staff before our students because it is the only hope we have of securing what our students need most: a world-class education.

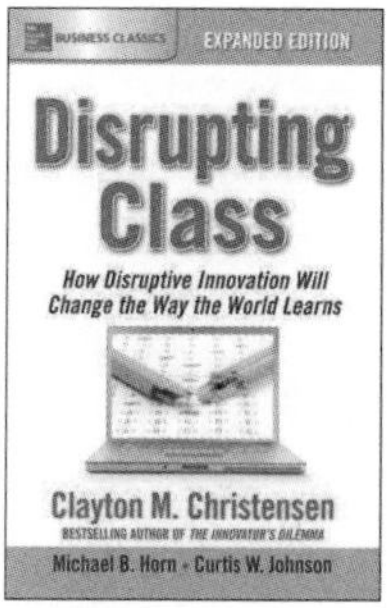

Disrupting Class: How disruptive innovation will change the way the world learns, Clayton Christensen, Michael Horn and Curtis Johnson.

Summary – Recent studies in neuroscience reveal that the way we learn doesn't always match up with the way we are taught. To stay competitive – academically, economically, and technologically – we need to rethink our understanding of intelligence and re-evaluate our educational system. *Disrupting Class* offers a ground-breaking and timely prescription for education reform that incorporates customized learning, student-centric classrooms, and new technology.

Thank you

Throughout this book, we've looked back at the early foundations of educational technology, the evolutionary process that has often been the dominant force, and at how poorly planned early adoption failures or short-sighted funding models have generated a justified cautiousness when hearing the battle cry to embrace EdTech.

Hopefully, the word 'EdTech' now appears larger than life for you through a wider lens, isn't restricted solely to classroom pedagogy and sparks your imagination and inquisitiveness to look beyond discussions about laptop vs tablet or local vs cloud to the need to align voices and bring all the possible components of EdTech together as part of a long-term plan.

I'll continue to remind everyone that at the heart of any strategy, project or plan are the people. Communicate well, support well, co-produce well and often, and the rest of the challenges will become far more surmountable. It's great to involve as many stakeholders as possible, but we can't keep adding to the expectations of our teams without freeing time up somewhere else, ideally freeing it up to be left as just that – free time. Technology and some of the solutions discussed throughout the book can be one way of starting to preserve the precious commodity of time and push the wellbeing balance back in the right direction. Hopefully, some of the ideas and approaches I've shared will help you to avoid the inevitable pitfalls and arm you with a list of questions to ask and ideas to explore.

EdTech is so diverse, and believe me, that's a good thing. In fact, every year, the scale and diversity of EdTech grows: the breadth of AR/VR resources is expanding exponentially; AI and (within that) machine learning offers huge opportunities for personalised learning, parallel learning to supplement teacher-to-student activities and the potential to broaden our perspectives and how we engage with EdTech. Perhaps for our digital interface of the future, we need to talk more and type less?

With diversity and choice comes challenge – picking the right solutions from an ever-growing shelf of EdTech options reminds us that the value of choice is only realised if, in parallel, we have the value of peer reviews, case studies, research-informed evidence and well-planned and structured evaluation cycles. Now, more than ever, quality must shine over quantity.

Like our children, no two schools are the same, and so I've tried to share the questions alongside the context, so you are better informed and able to move forward. But just like our unique children, there are always common factors that apply to all, and where that's the case, I've tried to share suggestions and links to suitable trusted resources.

Finally, looking through the widest lens of all, think about our education system: how we could achieve more with a long-term vision and plan for education and committed long-term funding to facilitate effective strategic planning and investment by schools; reflecting on and (where needed) enhancing our approaches to teaching, learning and assessment; prioritising learner-centred design; empowering our children to lead the charge as digital citizens; developing and connecting our students' onward paths into the workplace and, hopefully, a lifelong learning journey.

If I have encouraged you to think more about the future digital landscape in our schools, to ask more questions, reflect, take some risks and try some of the approaches and tools shared – then it has absolutely been worth the effort.

Most of all – genuinely – thanks for sticking with me to the last page. It's my first book and I have tried to share as much as I can. If you have found it helpful, enjoyable or a catalyst for fresh ideas, then please do share online and tag me in (@AlKingsley_edu), and if you haven't, then just like my diary, let's keep it our secret... 😉

Bibliography

Bellow, A. (2013) Closing keynote at ISTE 2013 – www.bit.ly/3ccAMdl

Bilton, N. (2012) Project Glass – *The New York Times* – www.nyti.ms/3uDwVNe

Caukin, N., Trail, L. and Hover, A. (2020) How EdTech can support social and emotional learning at School and at home, *International Journal of the Whole Child* 5 (1) – www.bit.ly/3uK7FVt

Children's Internet Protection Act (2000) – www.bit.ly/3g4ZkGx

Christensen, C. M. and Raynor, M. E. (2003) *The Innovator's Solution: creating and sustaining successful growth*, p. 215.

Cohen, P. A., Kulik, J. A. and Kulik, C. C. (1982) 'Educational outcomes of tutoring: a meta-analysis of findings', *American Educational Research Journal* 19 (2) pp. 237–248.

Couros, G. (2015) *The Innovator's Mindset: Empower Learning, Unleash Talent, and Lead a Culture of Creativity.*

Department for Education (2021) KCSIE, Keeping Children Safe in Education – www.bit.ly/3ihFN8R

Dweck, C. (2007) *Mindset: The New Psychology of Success.*

Education Endowment Foundation (2020) Remote Schooling report – www.bit.ly/3wVDQ63

Elias, M. J. (2016) How to Implement Social and Emotional Learning at Your School – https://edut.to/3c7o9Rd

Ellson, D. G., Barber, L., Engle, T. L. and Kampwerth, L. (1965) Programed tutoring: A teaching aid and a research tool, *Reading Research Quarterly* 1 (1) pp. 77–127. www.bit.ly/3pbSfZ9

Enser, M. (2021) Covid closures: how teachers adapted to working remotely, *Tes* – www.bit.ly/3cdXAd3

FELTAG (2014) Paths forward to a digital future for Further Education and Skills – www.bit.ly/3ia1LdN

Home Office (2021) Prevent Duty, Counter-Terrorism and Security act, 2015 – www.bit.ly/3gjoLof

JISC (2020) Teaching staff digital experience insights survey, Nov 2020 – digitalinsights.jisc.ac.uk

Kington and Mleczko (no date) *The Advantages of Successful School-Community Relationships: Findings from the Include-Ed project* – www.bit.ly/3uJ9Vwj

Kotter, J. P. (2012) *Leading Change.*

Kubler-Ross, E. (1969) *On Death and Dying.* New York: Simon & Schuster.

Lawless, K. A. and Pellegrino, J. W. (2007) Professional Development in Integrating Technology Into Teaching and Learning: Knowns, Unknowns, and Ways to Pursue Better Questions and Answers, *Review of Educational Research* – www.bit.ly/3vJ5tyQ

Nelson, M., Voithofer, R. and Cheng, S. (2019) Mediating factors that influence the technology integration practices of teacher educators, *Computers & Education* 128 – www.bit.ly/34DnUcz

OECD (2020) Back to the Future of Education – www.bit.ly/3g5lDMm

Ofsted (2019) Education inspection framework – www.bit.ly/3on8aDp

Pajaron, T. (2013) 30 ways Google Glass in education might work – www.bit.ly/3plwm9S

Pressey, S. L. (1926) *Introduction to the Use of Standard Tests*, pp. 373–376.

Puentedura, R. R. (2010) SAMR Model – www.bit.ly/3fFFKSu

Ranger, S. (2015) Windows RT: The odd birth, brief life, and quiet death of Microsoft's ugly duckling operating system – www.zd.net/2SLrleq

Researchandmarkets.com (2021) Europe EdTech and Smart Classroom Market Forecast to 2027: Coming Together of Latest Technologies for Enhanced Learning – https://bwnews.pr/3yW4k9d

Sandu, D. (2021) Show me the evidence: How do you procure your school's ed-tech? – www.bit.ly/3i8OVy9

Schleicher (2021) What will education look like in 20 years? – www.bit.ly/3wS9dOU

Seymour, B. (2021) Trust the Process: How to Choose and Use EdTech That Actually Works – www.bit.ly/3yYFVQv

Terada, Y. (2020) A Powerful Model for Understanding Good Tech Integration – https://edut.to/34G5LLe

Tondeur, J., Aesaert, K., Prestridge, S. and Consuegra, E. (2018) A multilevel analysis of what matters in the training of pre-service teacher's ICT competencies – www.bit.ly/3wUdZLs

Toro, J. M. de (2016) Five steps to develop a new product – www.bit.ly/2T5MaS1

Williamson, B. (2021) Meta-edtech, *Learning, Media and Technology* 46 (1) – www.bit.ly/2S5izb1

Other resources I have mentioned in this book that I'd recommend you check out

3 Big Questions schools should ask EdTech vendors, eSchoolnews March 2021 – www.bit.ly/2SSyVDT

50+ popular hashtags for educators, TE@CH with ICT – www.bit.ly/3vJgQa0

Academies Act 2010 – www.bit.ly/3uE0T3C

School Capital funding allocation 2020–2021 – www.bit.ly/2Reje9J

Al Kingsley: personal website – www.AlKingsley.com

American SPCC – www.americanspcss.org

Amplify (a complete digital classroom) – www.bit.ly/3pcefD0

ARCnet, embedded real-time network – www.arcnet.cc

Association for Learning Technology (ALT) – www.alt.ac.uk

Associations Between Time Spent Using Social Media and Internalizing and Externalizing Problems among U.S. Youth. JAMA Psychiatry, K.E.Riehm, K.A.Feder, & K.N Tormohlen, 2019.

Becta – www.bit.ly/3yQyvi9

Besa – www.besa.org

Bett Show (UK) – www.bettshow.com

British International School, Abu Dhabi – www.bit.ly/3wUC3xN

CASEL, The Collaborative for Academic, Social, and Emotional Learning – www.casel.org

Century AI personalised learning – www.century.tech

Child Mind Institute – www.childmind.org

Childnet International, online grooming – www.Childnet.com

Classroom.cloud: Classroom Management platform for in school and remote learning – www.classroom.cloud

Classroom Dynamics: The Impact of a Technology-Based Curriculum Innovation on Teaching and Learning, E.B.Mandinach and .H.F.Cline (Jan 1996) – www.bit.ly/3iaezRp

ClubHouse app – www.JoinClubHouse.com
Connect Safely, safety, privacy, security and wellness – www.connectsafely.org
CPOMS, safeguarding concerns – www.cpoms.co.uk
Creative Engineering: Promoting Innovation by Thinking Differently, Arnold, John E. 2016 – www.stanford.io/3g5BSsO
Cyber Essentials Scheme – www.bit.ly/2RXOFHK
Deira International Schools, Dubai – www.disdubai.ae
Design Thinking – www.bit.ly/3yZp2Vw
Digi Social – Digital life training platform for children – www.digiisocial.com
Digital Citizenship: Addressing Appropriate Technology Behaviour, Ribble, M.S., Bailey, G.D., and Ross, T.W. 2004 – Learning & Leading with Technology.
Digital Citizenship Institute – www.DigCitInstitute.com
Digital Content Evaluation & Selection Process checklist, Pickerington LSD – www.bit.ly/2SSWRHo
Digital signage for Schools, Trilby TV – www.TrilbyTv.co.uk
Distance Learning profiles of Dubai Private Schools – www.bit.ly/3wSZt6N
Distance learning: Reflections on the EEF's rapid evidence review, Dr Caroline Creaby, April 2020 – www.bit.ly/3peyQ9K
District Administration Magazine – www.districtadministration.com
Dogsthorpe Infants School co-production of ReallySchool – www.dogsthorpeinfants.co.uk
Douglas Hofstadter's Fluid Concepts and Creative Analogies: Computer Models of the Fundamental Mechanisms of Thought – www.bit.ly/3yWnxYhs
EdTech Demonstrator programme – www.bit.ly/3yTMOCF
EdTech Evidence exchange (US) – www.edtechevidence.org
EdTech evidence Group – www.edtechevidencegroup.com
EdTech Impact (UK) – www.edtechimpact.com
Education Alliance Finland – www.educationalliancefinland.com
Education Endowment Foundation – Remote Learning, Rapid Evidence Assessment, London: Education Endowment Foundation. Accessed 14 August 2020 from www.bit.ly/3x7Ol2D
EduGeek – www.Edugeek.net
Effective Communication, Leah Davis – www.bit.ly/3yneCPx
Electronic Classroom of Tomorrow (ECOT) failure – www.bit.ly/3x7ONDt

ESafety, Keyword Monitoring – www.NetSupportDNA.com/education
eSchool News Magazine – www.eschoolnews.com
Essential elements of digital citizenship, M.Ribble, 2020 – www.bit.ly/3g9H4vV
Everyone's Invited – www.everyonesinvited.uk
F6S – www.F6S.com
Foundation for Education Development (FED) – www.Fed.education
GESS show (UAE) – www.gessdubai.com
Get Help with Technology programme, DfE – www.bit.ly/3vNsBwh
Global Cloud Infrastructure market 2020 – Canalys data – www.bit.ly/3hJztGz
Global EdTech, International EdTech news – www.global-edtech.com
Global Venture Capital in EdTech 2020, Holon IQ – www.bit.ly/34JiAUY
GoBubble, A safer, healthier and kinder digital community forschools – www.GoBubble.school
GroupCall Messenger – www.groupcall.com/systems-for-schools
HeadTeacher Update magazine – www.headteacher-update.com
The EdTech Evidence Group – www.edtechevidence.com
How to Teach Your Students the 9 Elements of Digital Citizenship, 2019 – www.bit.ly/3fJXGLv
ICT in UK Maintained Schools 2019, BESA – www.bit.ly/2RXaw09
Internet Matters, Helping parents keep their children safe online – www.internetmatters.org
Internet Watch Foundation – www.iwf.org.uk
ISC International School Awards – www.bit.ly/34LocxQ
ISTE Show (US) – www.ISTE.org
JESS (Jumeirah English Speaking School) Digital Innovation Summit – www.bit.ly/34Jr7qQ
Judge Business School, Cambridge University – www.jbs.cam.ac.uk
Kano model – Categorising customer requirements – www.kanomodel.com
Learn Ed – EdTech roadshows and resources – www.LearnEd.Org.uk
Linda Parsons – www.digilinlearning.com
Lippitt-Knoster model for managing change – Mary Lippitt & Timothy Knoster, 1987
Mark Anderson – ICT Evangelist – www.ictevangelist.com
Mentimeter, Live interactive Polls & Quizzes – www.Mentimeter.com
MyConcern, safeguarding concerns – www.myconcern.education

National Bullying Helpline – www.nationalbullyinghelpline.co.uk
National Children's Alliance – www.nationalchildrensalliance.com
Nesta: The innovation foundation – www.netsa.org.uk
NetSupport: award-winning developer of software solutions for education – www.NetSupportSoftware.com
NSPCC – www.NSPCC.org.uk
Ofsted, Office for Standards in Education, Children's Services and Skills – www.gov.uk/government/organisations/ofsted
Olly Lewis, Digital Strategy – www.OllyLewisLearning.com
Parent Apps – www.parentapps.co.uk
ParentMail – www.parentmail.co.uk
Parents' Evening Manager – www.iris.co.uk
Philippa Wraithmell, Designed to Teach Digitally – www.designedtoteachdigitally.com
RCPCH (2019). The health impacts of screen time – a guide for clinicians and parents. Accessed 14 August 2020 from www.bit.ly/3wSDEUX
ReallySchool – Classroom Observations, Tracking and learning journals – www.ReallySchool.com
Reflecting on remote teaching: Exemplifying several strategies, Steve Smith May 2020 – www.bit.ly/3pseZUZ
Remote Learning Solutions: Crowd-sourced ideas for checking students' writing, Tom Sherrington – www.bit.ly/3uGnigM
Rosendale Primary School – www.rosendale.cc
Royal Society for Public Health, Safe Social Media – www.rsph.org.uk
Russell Prue – School Radio solutions and Training – www.Andertontiger.com
School Booking – www.schoolbooking.com
SchoolCloud – www.parentseveningsystem.co.uk
SchoolComms – www.schoolcomms.com
SchoolTrustee.blog – Resources for School Governors & Trustees – www.SchoolTrustee.blog
Secondary School ICT budgets drop by £17m, Education Technology Magazine, August 2019 – www.bit.ly/3uGnmx2
Stop Bullying campaign – www.stopbullying.gov
Teachers and teacher education in uncertain times, Madalińska-Michalak et al. 2018 – www.bit.ly/2SXat4E

Technological Pedagogical Content Knowledge framework, P.Mishra and M.J. Koehler 2006 – www.bit.ly/34GeHjJ
The Association of Network Managers in Education – www.ANME.co.uk
The EdTech Genome Project, 2019 – www.edtechevidence.org/edtech-genome-project
The Education Foundation, promoting use of EdTech in the UK – www.ednfoundation.org
The Innovator's Dilemma – Clayton Christensen, 2013
The International Society for Technology in Education (ISTE) – www.ISTE.org
Truth for Teachers podcast, Angela Watson – www.bit.ly/3g6uGN9
UCL Institute for Education – www.ucl.ac.uk/ioe
UK 'Free Schools' programme – www.gov.uk/types-of-school/free-schools
UK Department for Education – www.bit.ly/3eIfpTk
UK Safer Internet Centre – www.saferinternet.org.uk
US Charter Schools Program – www.bit.ly/2RsHd4Q
What your students really need to know about Digital Citizenship, Vicki Davis, Edutopia Nov 2019 – https://edut.to/3wTqOkC
Wikipedia – Educational Technology definition – www.bit.ly/2RXjISd
Worldwide Public Cloud spend 2021 – Gartner – https://gtnr.it/3bIGL9L
Year 2000 problem – Wikipedia – www.bit.ly/3yU3Hx3

I'm hugely proud of the team and solutions at NetSupport. If you would like to find out more, please scan the QR code below to visit our education solutions showcase.

NetSupport

www.NetSupportSoftware.com/education-solutions